Yew-Jin Lee • Jason Tan
Editors

Primary Science Education in East Asia

A Critical Comparison of Systems and Strategies

Springer

Editors
Yew-Jin Lee
National Institute of Education
Nanyang Technological University
Singapore, Singapore

Jason Tan
National Institute of Education
Nanyang Technological University
Singapore, Singapore

ISSN 1878-0482 ISSN 1878-0784 (electronic)
Contemporary Trends and Issues in Science Education
ISBN 978-3-030-07313-8 ISBN 978-3-319-97167-4 (eBook)
https://doi.org/10.1007/978-3-319-97167-4

This Springer imprint is published by the registered company Springer Nature Switzerland AG
The registered company address is: Gewerbestrasse 11, 6330 Cham, Switzerland

Contemporary Trends and Issues in Science Education

Volume 47

SCOPE

The book series Contemporary Trends and Issues in Science Education provides a forum for innovative trends and issues connected to science education. Scholarship that focuses on advancing new visions, understanding, and is at the forefront of the field is found in this series. Accordingly, authoritative works based on empirical research and writings from disciplines external to science education, including historical, philosophical, psychological and sociological traditions, are represented here.

More information about this series at http://www.springer.com/series/6512

Contents

About the Authors

Yu Chen is a senior research assistant in the Department of Science and Environmental Studies at the Education University of Hong Kong. Her research focuses on socioscientific issue-based instruction, technology integration, and primary school science education.

Teng-Yao Cheng is currently a PhD student of the Institute of Network Learning Technology at National Central University. Teng-Yao had taught as a science teacher. His research interests include science education, network learning, massive open online course (MOOC), scientific reasoning and argumentation, technology-enhanced science learning, and knowledge-building pedagogies.

Esther G. S. Daniel is attached to the Department of Mathematics and Science Education, Faculty of Education, University of Malaya. She has been a teacher educator for more than two decades. Her areas of expertise and background are in science education, environmental education, technology integration in the classroom, and learning and cognition. Esther has conducted numerous training workshops and seminars related to research, pedagogies, and technology integration in education. Her research has focused on science education, in particular, environmental education. She has published papers in journals as well as presented at various conferences both local and international. She has written numerous training modules for the teaching and learning of science and for environmental education. More recently, she has also been involved in medical education with a focus on the psychology of learning. Her latest book is entitled *Biology Education in a Changing Planet*. In 2016, Esther received the UNESCO-HAMDAN international award for the effective training of teachers in Paris from the Director General of UNESCO on October 5, 2016, which is the day designated as World Teachers' Day. More recently, she received the Venus International Women's Award (VIWA) for a "Lifetime Achievement in Science Education (Specialisation-Biology)."

Weiping Hu received a Bachelor of Science in physics from Shaanxi Normal University in 1984, a master's degree in physics education from Beijing Normal University in 1998, and a PhD in educational psychology from Beijing Normal University and King's College London in 2001. Now, he is the director and professor of the Key Laboratory of Modern Teaching Technology, Ministry of Education (China), Shaanxi Normal University, vice director of the Collaborative Innovation Center of Assessment toward Basic Education Quality (China), and president of the National Association for Science Education, the Chinese Society of Education. His research interests include developmental and educational psychology, science teaching psychology, and STEM education.

Mijung Kim is an associate professor in science education at the Faculty of Education, University of Alberta, Canada. Her research interests include science inquiry, dialogical argumentation, and children's collective reasoning and problem-solving in science classrooms. Her current publications include journal articles on inquiry-based teaching, collaborative problem-solving, and decision-making on socioscientific issues, and book editions, *Biology Education for Social and Sustainable Development* (M. Kim & H.C. Diong, 2012, Sense Publisher) and *Issues and Challenges in Science Education Research: Moving Forward* (D. Tan & M. Kim, 2012, Springer). She has been currently involved in a UNESCO project, "International guidebook on embedding concepts of peace, sustainable development and global citizenship in textbooks of core subjects," as one of science subject experts in 2016–2017.

Mee-Kyeong Lee is the director of the Curriculum and Textbook Division at the Korea Institute for the Curriculum and Evaluation (KICE), a leading research institute at the national level dedicated to improving the quality of education by conducting a wide range of research on curriculum and evaluation in elementary and secondary schools. Dr. Lee participated as the national project manager for both PISA and TIMSS. She has also implemented various researches on National Science Curriculum Development, National Assessment of Educational Achievement (NAEA), and teaching and learning in science. She is currently doing research on classroom assessment and participating in the OECD Education 2030 project.

Yew-Jin Lee was formerly trained as a secondary school biology teacher in Singapore. His current interests are in primary science, curriculum research, scientific ways of knowing, and informal learning environments as well as classroom assessment. Overall, he tries to understand how people learn and under what conditions does effective and powerful learning take place. He was the past coeditor of *Pedagogies: An International Journal* (Routledge) and serves on the editorial boards of *Research in Science Education*, *Studies in Science Education*, and *Asia-Pacific Science Education*. In 2015, a co-authored review on cultural-historical activity theory in education (Roth & Lee, 2007) achieved a highly cited SSCI article status (top 1%) within the discipline of the social sciences.

Adam Lefstein is associate professor in education at the Ben-Gurion University of the Negev in Israel, where he conducts research and teaches about pedagogy, classroom interaction, teacher learning, and educational change. He is particularly interested in the intersection between research and professional practice and how to conduct research that is meaningful, rigorous, and helpful for educators. Among other activities, he directs the Laboratory for the Study of Pedagogy, an interdisciplinary research group that is committed to rigorous investigation of Israeli schooling, pedagogy, and educational policy; the development of innovative research tools for the study of these phenomena; and processes of knowledge sharing with education practitioners, policy-makers, and the public. Currently, he is conducting a large design-based implementation study of teacher leadership and professional discourse. His book (with Julia Snell), *Better than Best Practice: Developing Teaching and Learning Through Dialogue*, was published in 2014 by Routledge (http://dialogicpedagogy.com/).

Kenji Matsubara is a senior researcher in science education at the Department for Curriculum Research, Curriculum Research Center, National Institute for Educational Policy Research (NIER), Japan. He worked as a science teacher at a lower secondary school and as a physics teacher at an upper secondary school in Japan, as well as a volunteer teacher in Zambia. After having earned his PhD in education at Hiroshima University, he implemented TIMSS 2011 and TIMSS 2015 as one of the National Research Coordinators at NIER. He has also been a project member of PISA and TALIS as well as other project researches such as Study on the Curriculum to Nurture Competencies by NIER. His research interests include science curriculum development, lesson analysis, cross-curricular teaching and learning such as STEM, and international educational cooperation. He is currently a member of the board of directors for the Society of Japan Science Teaching (SJST).

Elizabeth (Betsy) McEneaney is an associate professor in the Department of Teacher Education and Curriculum Studies at the University of Massachusetts (UMass) Amherst. A former high school mathematics and chemistry teacher, she earned a PhD in sociology at Stanford University with a focus on the globalization of ideas about primary school science and math curriculum. Her current research interests include equity in STEM fields, engineering education in PK-12, the impact of ubiquitous ICT on the curriculum, the dynamics of small group work, and research methods. She serves as the coordinator of the PhD program in Teacher Education and School Improvement at UMass and is an associate editor for the *Journal of Curriculum Studies*. The American Educational Research Association recently recognized her as an "Outstanding Reviewer," and she has received a number of teaching awards.

Umesh Ramnarain is a professor in science education at the University of Johannesburg in South Africa. His research is on inquiry-based science education, with a particular focus on its uptake in South African classrooms, where the unequal funding policies of the previous Apartheid education system have resulted in learning contexts that are complex and diverse in terms of physical resources, the educational and cultural backgrounds of learners and teachers, and school ethos. The importance of his work has been recognized internationally. His work has been published in top tier journals such as *Journal of Research in Science Teaching*, *International Journal of Science Education*, *Research in Education*, *Teaching and Teacher Education*, and *Journal of Curriculum Studies*. He has also disseminated his work at NARST and ESERA conferences. He is the associate editor of the international journal, *Research in Science Education*, and a member of the editorial board of *Journal of Research in Science Teaching*. He has received best paper awards at international conferences such as International Science Education Conference in Singapore and International Conference for Science Educators and Teachers in Thailand.

Terry Russell is professor emeritus of the University of Liverpool. His interest in developmental psychology originated during undergraduate studies at the University of Manchester. In particular, the challenge of understanding how we each construct knowledge of the world stimulated a lifelong interest in developmental psychology applied to education. After some experience of teaching primary and secondary pupils, Terry gained a professional qualification in educational psychology from Sussex University. A strong inclination to work in research was realized via a career shift with the ASEAN Regional Centre for Science and Mathematics, in Malaysia, supporting science concept development research. Returning to the UK, he took up a research post with the Centre for Science and Mathematics Education (CSME) at King's College London, London, working on the national assessment of science. He completed his PhD at the University of Liverpool where he directed the Centre for Research in Primary Science and Technology for three decades. Research and development projects were managed in national assessment and various aspects of science understanding, mostly with the age ranging from 3 to 14, including international development projects. His current research and writing is focused on developmental progression, specifically in relation to pupils' understanding of evolution.

Xin Shou is a doctoral candidate at Shaanxi Normal University majoring in science education. He received a Bachelor of Science Education from Chongqing Normal University in 2010 and a master's degree in science curriculum and instruction from Chongqing Normal University in 2013.

Wing Mui Winnie So is a professor of the Department of Science and Environmental Studies at the Education University of Hong Kong, the director of the Centre for Education in Environmental Sustainability, and the associate dean of the Graduate School. Her main research areas are STEM education, inquiry learning in science and environmental education, and integrating information technology and teacher development in science/general studies/liberal studies. She has been active with professional contribution, being the past president of the Asia-Pacific Education Research Association, executive member of the World Education Research Association, Hong Kong Educational Research Association, and East-Asian Association for Science Education, respectively.

Lorraine Pe Symaco is professor under the ZJU 100 Programme at Zhejiang University (ZJU), China. She was the founding director of the Centre for Research in International and Comparative Education (CRICE) at the University of Malaya, Malaysia. She was also UNESCO Chair in International and Comparative Educational Research with Special Reference to South East Asia.

Jason Tan is associate professor in policy and leadership studies at the National Institute of Education, Singapore. He completed his doctoral studies in comparative education at the State University of New York at Buffalo. His research interest includes education reform. Among his coedited books are *Going to School in East Asia*, *Challenges Facing the Singapore Education System Today*, and *Education in Singapore: Taking Stock, Looking Forward.*

Zhi Hong Wan is an assistant professor in the Department of Curriculum and Instruction at the Education University of Hong Kong. Before starting his research in science education, he had taught middle school physics for 5 years. His current research interests include nature of science, science learning, and higher-order thinking. He has published papers in a number of international journals, including *Science Education*, *Studies in Science Education*, *International Journal of Science Education*, *Research in Science Education*, *Science & Education*, *Thinking Skills and Creativity*, *Teaching in Higher Education*, *Science Education International*, and *Asia-Pacific Forum on Science Learning and Teaching.*

Li-Jen Wang is currently a full-time lecturer of the Language Teaching and Research Center at National Chiao Tung University and also a PhD student of the Institution of Network Learning Technology at National Central University. Li-Jen was educated at the University of Stirling, Scotland, UK, and graduated with MSc in Teaching English as a Foreign Language (TEFL). After Li-Jen started his PhD program, he also found his interests in science education. His research works involve not only language teaching and learning but science education. His research interests include language learning and teaching, computer-assisted language learning (CALL), English as an international language (EIL), game-based learning, science education, scientific reasoning and argumentation, and knowledge-building pedagogies.

Ying-Tien Wu received his PhD degree in the Department of Earth Sciences (science education) from National Taiwan Normal University (NTNU). He is now an associate professor of the Graduate Institute of Network Learning Technology. His research work involves both science education and educational technology, and his research interests include inquiry-based learning, scientific reasoning and argumentation, knowledge-building pedagogies, technology-enhanced science learning, and technological pedagogical content knowledge (TPACK). Ying-Tien has primary science teaching experience which benefits him for transforming educational theories into practice. He is keen to help science teachers to improve their teaching practices. His research findings have also been published in reputable international journals in science education and digital learning. He also serves as the editorial board member or the reviewer for more than ten influential academic journals in science education and digital learning (including *SSCI* journals). Owing to his outstanding research performance, he received the Ta-You Wu Memorial Award (Young Outstanding Researcher Award) from the National Science Council, Taiwan, in 2008.

Zong-Rong Yang is currently a science teacher at Taichung Wong-Zih Elementary School and also a PhD student of science education at National Changhua University of Education. His research interests include inquiry-based learning and STEM education. Zong-Rong has abundant science teaching and relevant experiences. Besides science teaching, he also devotes himself to various science education activities such as instructor in a science club, judge of science fairs locally and nationally, a lecturer at the National Natural Science Museum, and a primary investigator of several national research projects. Because of this devotion to science education, Zong-Rong also received the Best Instructor of National Science Fair Award in 2014 and the Distinguished Science Educator Award of Taichung City in 2014.

Hye-Gyoung Yoon is a professor in science education at Chuncheon National University of Education, which is specialized for elementary teacher education in South Korea. She has been interested in how to develop elementary preservice teachers' expertise on science inquiry teaching through collaborative reflection. She has been dedicated to developing science curriculum and practice in classroom levels in Korea. Currently, she participated in the development of national elementary science textbooks and Science Curriculum 2015. Her recent research interests and publications include the topics of "using visual representations in science classrooms" and "teachers' professional vision development through video-based reflection."

Carla Zembal-Saul, PhD, is a science educator, science teacher educator, and elementary educator. She holds the Kahn Endowed Professorship in STEM Education at the Pennsylvania State University. Her work is situated in school–university–community partnerships in the USA and abroad. Zembal-Saul's research investigates instructional practices and tools that support preservice and in-service elementary teachers in engaging children productively in scientific and engineering practices, with an emphasis on argumentation and evidence-based explanation construction. She is deeply invested in practitioner inquiry and video analysis of teaching as mechanisms for advancing teacher learning and development across the professional continuum. In addition to contributing to the research community, Zembal-Saul is committed to collaborating with teachers, bridging research and practice, and co-authoring publications with practitioners. She was recognized as a National Science Teachers Association Fellow in 2015 and served on the National Academies of Sciences, Engineering and Medicine: Board on Science Education consensus committee that authored the report, *Science Teachers' Learning: Enhancing Opportunities, Creating Supporting Contexts* (2015).

Chapter 1
Introduction

Yew-Jin Lee and Jason Tan

International or widespread interest in East-Asian education systems has boomed over the past two decades. (Re)beginning in the mid-1990s, researchers were asking questions about the complex but tantalizing causal relationships between educational investments in science education and national economic growth. Much to the disappointment of policymakers everywhere, Benavot (2002, p. 61) argued that the "link between the degree of industrialization or economic development of a nation and its curricular emphases is rather weak." As a case in point, although five identified developing countries in 1970 had a fairly large proportion of students engaged in science education, of these, only South Korea, Hong Kong, and Singapore experienced significant increases in GNP per capita, whereas the two remaining Latin American countries witnessed modest levels of growth. A number of reasons were offered to account for this unanticipated disparity in outcomes inter alia poor technology transfer, governance problems, inadequate provision for basic education, and unique sociocultural factors (see Aldcroft 2000; Ashton et al. 2000).

Another development puzzle involved the then much-vaunted Asian Tiger economies of Hong Kong, Singapore, South Korea, Taiwan, and Thailand. It was found that *before* these regions became economic powerhouses, exposure to instruction in school science here was relatively low (Benavot 1992) just as lessons were typically taught using vernacular languages and with strongly teacher-centered methods (King 1989). The latter, of course, gives few insights as to the actual quality of teaching in the subject, but it posed a challenge to those seeking direct relationships between education and the economy. Similarly, the wealth of East-European countries was believed to have been affected more by wider macroeconomic events and political upheavals rather than whatever heavy investments they had earlier devoted to science education (Sachs 2005). It soon became clear that regional- and/or

Y.-J. Lee (✉) · J. Tan
National Institute of Education, Nanyang Technological University, Singapore, Singapore
e-mail: yewjin.lee@nie.edu.sg

Y.-J. Lee, J. Tan (eds.), *Primary Science Education in East Asia*, Contemporary Trends and Issues in Science Education 47,
https://doi.org/10.1007/978-3-319-97167-4_1

country-level differences confounded making any specific statements regarding the drivers of human capital development (Drori 1993), while a number of methodological and theoretical problems plagued analyses of the available data (Drori 2000). It was, nonetheless, acknowledged by most researchers that investment in science (education) was a necessary but not sufficient condition for economic progress (Caillods et al. 1997).

The second major and interrelated reason for the intense scrutiny that some East-Asian states were facing was due to the stellar performances of students from regions such as South Korea, Mainland China, and Hong Kong in international comparative education assessment exercises, namely, the Programme for International Student Assessment (PISA) and Trends in International Mathematics and Science Study (TIMSS). These achievement tests have now come to play a normative, almost hegemonic role during national educational policy discourses (Pereyra et al. 2011; see also Rautalin et al. 2018); politicians and educationists in Australia, England, and the United States among others have made explicit references to East-Asian educational successes as part of their politically tinged rhetoric with regard to educational reform (Darling-Hammond et al. 2017; You and Morris 2016).

That there exists a mystique regarding the academic prowess of students and their teachers/textbooks/curriculum in East-Asian science, technology, engineering, and mathematics (STEM) subjects is surely an understatement. In lands far from Asia, there is a growing willingness if not impatience to emulate hitherto unfamiliar curriculum content, pedagogical strategies, and teacher training policies from these allegedly superior systems as part of policy transfer and borrowing (Forrestier and Crossley 2015). At least 14 jurisdictions from as different as South Africa and Chile, for example, have been enthusiastic consumers of "Singapore math." This way of teaching has been reported to raise test scores because it is based on a model method of instruction that in turn was inspired by ideas about human learning from Jerome Bruner (Yuen 2017). China, too, has been an exporter of mathematics instruction to the West given the intellectual rigor and success of teaching this difficult subject in that country. However, various education policy researchers have also cautioned against the all-too-prevalent tendency to ignore the peculiarities of sociocultural and institutional contexts in the search for magic bullet policy prescriptions even while popular stereotypes about East-Asian students and education systems continue to enchant so many (Takayama 2017). To complicate matters, multilevel/hierarchical modeling of PISA scores also suggests that socioeconomic and demographic aspects of test-takers exert stronger influences than any changes in national policies such as curriculum reforms, an old leitmotif among critical sociologists of education (Aloisi and Tymms 2017).

This edited volume therefore steps right into this policy context to fill a much-needed void, namely, that of an authoritative description and critical comparative analysis of the structure and organization of primary science education in six East-Asian regions—the People's Republic of China, Republic of Korea, Republic of China, Hong Kong SAR, Japan, and Singapore. Each of the country- or region-focused chapters has been authored by active researchers or practitioners with an intimate knowledge of the ways in which national history, politics, economics, and culture have contingently affected the development and evolution of their elementary science systems.

Equally important are the six invited commentary chapters that cast a discerning eye over the major issues, conundrums, and blind spots in East-Asian primary science education. These types of synoptic overviews add much value: enabling others to understand broad commonalities and helping to synthesize what must surely be a bewildering amount of very interesting yet often contradictory body of reforms and policies. East Asia indeed has many lessons (both positive and negative) to offer to the rest of the world, although few scholarly resources currently provide reliable comparisons on specific areas of interest (e.g., the organization or frameworks of elementary science) that are written in English.

In a nutshell, we hope that this tome will provide readers with a state-of-the-art report regarding the national systems and strategies of primary science education in East Asia. Readers would have an accurate sense of what has been accomplished, what has not worked so well, and what remains to be done. Far from adopting a naïve "best practices" approach, these chapters provide readers with clues that may serve as the basis of future reforms for enhancing science education both within and outside of the East-Asian region. This is our wish as co-editors where one is a science educator and former high school teacher (YJL) and the other a comparative education researcher cum curriculum policy expert (JT).

1.1 Why East Asia, and Why Primary Science?

Speaking with respect to primary education in the developing world, policy analysts Benavot and Kamens (1989, p. 3) once claimed that "with all the interest in providing an instructionally effective and financially efficient educational environment, it is surprising how little is said (or known) about one of the most important components of schooling in the modern world: the curriculum." We concur that after three decades, research on such fundamental, coordinating aspects of schooling in Asia are generally hard to come by, what more in science education. Scholarly publications or official reports are often written exclusively in vernacular tongues or were completed as small-scale projects commissioned by local governments that curtailed their diffusion to the wider community.

When we realize that more than half of the world's population live in Asia and yet we know so little about the educational systems here, this becomes a very regrettable situation. It comes as no surprise that science education originating from Asian contexts is not as well-publicized or understood as it ought to be (Lee 2008). We are therefore thankful that our knowledge of the former has been improving in recent years: Liang et al. (2017) in this same book series offered rare insights into science education from China, while Chiu (2016) performed a similar service with respect to Taiwan. Lin et al. (2016) produced an important collection of studies from East Asia that was paralleled in an earlier book by Khine (2015) and in a handbook by

Lee (2010). A special issue in 2007 highlighting science conceptual understanding research from Taiwan (Chiu et al. 2007) in the premier journal of the discipline, *International Journal of Science Education*, was perhaps a breakthrough for Asian science education. This was joined 5 years later by another special issue in the same journal now showcasing investigations from Mainland China (Liu et al. 2012). Much remains to be unpacked as catering to the Asian Chinese learner of science presents more than a few paradoxes that are as yet incompletely explained (Cheng and Wan 2015).

One interesting though telling reality that has emerged is that there is a direct correlation between the overall quantity and/or quality of educational research and regions that are wealthier or politically stable. It was, more than anything, for these pragmatic rationales concerning ease of access to existing research communities in science education that decided our selection of regional case studies from Asia. Compared to other parts of Asia, a number of East-Asian states such as those described in this book are therefore in a highly enviable position where basic provisioning and services in urban schools are already satisfied leaving their Ministries of Education to focus on raising quality in these privileged districts.

Yet for science educators, general information regarding the structure, systems, and strategies associated with primary science in East Asia remains very sketchy or piecemeal at best. This might have been a fallout of various long-standing debates in the West concerning what was appropriate for younger children: On one hand, there were doubts about the ability of young children to manage the demands of doing authentic science activities, to engage in "minds-on" science. On the other hand, there was also a fear of too early induction to secondary science that prevented teaching science in primary schools other than the amorphous and cognitively undemanding subject of "nature studies" (Atkin and Black 2003; Harlen 2014a, b).

It seems remarkable that a book that one of us (YJL) co-authored examining the conceptual demands of learning outcomes from elementary science curricula in East Asia was only recently released when such basic understandings should have been common knowledge a long time ago (Lee et al. 2017). Although we lack sufficient knowledge about the overall frameworks or implementation of primary science in Asian public education (see Kim et al. 2015 for an excellent overview of secondary science in East Asia), many other related questions haunt us such as what passes for inquiry science and student epistemic practices here, levels of student autonomy, assessment practices other than standardized tests, nature of classroom climate, frequency of dialogic interactions, and many other consequential domains that have been better investigated in Western contexts. Quite simply, we are unable to fathom what the everyday experiences of science are like for a child in a primary school in East Asia. Nor do we possess a good sense of what kinds of scientific literacies we are ultimately nurturing through school in East Asia: One that perhaps pays homage to future well-paying occupations or something that intellectually enriches and sustains one throughout adult life? Because research on the above has been so neglected or dispersed in the literature, we believe that our edited volume will be a timely consolidation.

We acknowledge that the enactment of science education is but a subset of the enterprise of public schooling in general; for example, the opportunities for learning in the subject are constrained by the allocated number of school hours and class sizes, factors which are completely beyond what ordinary science teachers can control. Furthermore, science teaching/learning is intertwined with larger educational goals or reforms in any country such as those promoting interdisciplinary curricula, distributed school leadership, or computer technologies. In this book, we make reference to these expansive though "external" structural or cultural forces that can be explanatory minefields. However, we concentrate on matters closer home to disciplinary teaching regarding curricula or syllabi, textbooks, pedagogies, assessment issues, laboratory work, science teacher preparation and professional development, out-of-school science, and the like. We cannot guarantee that the secrets of East-Asian "successes" in PISA or TIMSS will be self-evident after this publication (see Jerrim 2015; Lau and Lam 2017; Lee et al. 2015), but we can guarantee reducing some of the misconceptions surrounding the organization and conduct of elementary science in East-Asian schools.

What we do know, and which the diligent reader can quickly verify, is that schooling and curricula are experiencing convergence around the world (see McEneaney 2003): There is a distinct pattern or movement toward homogeneity of educational approaches and instruments such as standardized tests and nationally prescribed learning outcomes or objectives (Jenkins 2015; Loomis et al. 2008; Wilkinson 2013). Taking back, with great gusto if we may say what has traditionally been the professional responsibility of science departments and teachers—curriculum development—the Ministries of Education are now entrenched in their role as chief determiner of what gets taught and tested (Elyon 2014; Harlen 2014a). It goes without saying that in all the jurisdictions described here, teacher training is taken extremely seriously with candidates expected to obtain mastery of subject matter and contemporary teaching methods including upholding the correct moral values.

These examples of increased governmentality have been occurring in the West as well as in the East, although the latter has historically been more accepting of the supposed benefits arising from mass enrolments in science education and from the repeated recommendations by international aid agencies such as the World Bank, the Asian Development Bank, and UNESCO (United Nations Educational, Scientific, Cultural Organization) (see Harlen 1994). Apart from justifying such reform measures due to the greater need for accountability and better customization of learning, many national documents now espouse so-called twenty-first-century skills (Sinnema and Aitkin 2013). From a study of 11 high-achieving regions, it has also been reported that science curricula in Asia have been encouraging creativity and application of knowledge among students, whereas curricula in Western states were becoming more knowledge focused as well as prescriptive with regard to content and teaching (Hollins and Reiss 2016). While it is hard to deny the current logic of improving the life chances of young people through new skills including a solid grounding in STEM subjects, what surely cannot be known with any certainty is how different groups of people under different contexts stand to benefit or lose out—where you are *now* matters a great deal (Allen 2017).

In this book we definitely are comparing six systems of elementary science, but our driving impetus stems from our desire to learn from one another, to appreciate why alternative arrangements have developed in a particular manner. We also desire to understand some of the (expected and unexpected) policy consequences and above all to intensify our gaze at one's own system for fresh critique. Given how politicians have long capitalized on panics of various sorts to further certain agendas, we wish to make our personal stances visible:

> A strong temptation, however, would be to regard these findings as the latest photo-finish from an intense competition to confirm which state is the "best" in whatever way one would care to define that notorious label. We thoroughly reject such narrow interpretations of the data. (Lee et al. 2017, pp. 2–3)

We trust that our readers will apply a kind of refined connoisseurship when examining the nuances and possibilities of primary science education here, which is none other than understanding one's context better through a process of valuing the other (Manzon 2014).

1.2 A Road Map for Reading

1.2.1 Part I: The Regional Chapters

To begin in this important review task, invited scholars for each substantive chapter from the six East-Asian regions were encouraged to address, as far as reasonably possible, the following topical concerns of systems and strategies in elementary science as listed below:

1. Overview of local primary science education, perhaps even at the level of provinces/school districts if relevant or interesting
2. General educational and/or specific policies pertaining to education including primary science education
3. Curriculum visions, goals, emphases, and structure of primary science in that region
4. Pedagogy/teaching and other innovations that are prevalent or trending
5. Assessment and testing practices
6. Teacher learning and training
7. Recent reform attempts including failed reforms
8. Impact or influence of TIMSS, PISA, and other international rankings on the local educational landscape
9. Cross-national borrowing or adaptation of ideas. Local innovations unique to the state/territory/region

We do not wish to rehearse in detail what these chapters in Part I have already very competently accomplished as they introduced what primary science education looks like and the possible reasons why it evolved in this fashion in their respective regions. Each place is different, and hence our authors had the freedom to express what they felt was relevant or problematic and worthy of sharing. Given that these are in essence stand-alone chapters, readers can directly zoom into whichever region they wish. There is no correct sequence to be followed, although after reading them all, one might also be quite surprised at the variability of practices and policies to be found here.

Despite the threat of oversimplification, we can attempt a broad overview of the six chapters to pique our readers' interest: Every chapter began by explaining the history or organization of primary science education and/or the science curriculum by situating these within the educational milieu of their region. Many contributors' referenced policy changes that occurred after World War II and with good reason for every region here experienced immense human suffering and material damage during those years of conflict. Although not highlighted explicitly, one region is still technically at war (Korea), while another changed sovereignty two decades ago (Hong Kong). Another's independence has been in dispute (Taiwan) for years, and two others gained statehood only after the war (China, Singapore). These sociopolitical upheavals have affected their education systems as well as shaped the diverse purposes of schooling in their respective societies, which space limitations have prevented a fuller explication.

As was the case elsewhere, the initial focus of elementary science in East Asia was generally biased toward biology, human health, or nature studies (e.g., Hong Kong, Japan) that then shifted toward emphasizing process skills, hands-on work, and experimentation activities including that in the physical sciences. Such a historical progression from learning about everyday life experiences to engaging in inquiry and process skills followed by infusing science, technology, and society (STS) elements and now a concentration on interdisciplinary learning and creativity has in fact been described by the Korean contributors. That all these educational developments parallel the evolution of science teaching in the West is not unusual; Asian educators were exposed to and embraced what they felt was cutting-edge ideas about how science ought to be taught.

Nearly all regions can now boast of some form of ideologically coherent science curriculum frameworks thanks to multiple teams of disciplinary experts and teachers authorized and supported by local and state governments. Their concerted efforts have resulted in a slew of textbooks, teaching guides, workbooks, and assessment books that are an integral part of the pedagogic arsenal in East-Asian school systems. Many authors in this book have therefore shown typical examples of science workbooks from their own regions, a daily buffet of practice and disciplinary learning experienced by countless students here that is arguably less prevalent in the West. While some might legitimately criticize the rigidity and potential deskilling of teaching that might ensue, others might perhaps see some positives: Systems here provide comprehensive support for beginning teachers or for those who work in less advantaged areas. In other words, these educational materials or packages can compensate for uneven teacher quality, a massive problem in much of Asia (Asian Development Bank 2015). And all chapter contributors would quickly agree with the country report from Japan that even though their regions might follow a national curriculum, teachers here do have the freedom to decide how they teach for their diverse students.

This well-organized ecosystem for teaching-learning is guided by region-specific visions, aspirations, and purposes of primary science, although in more recent years, one observes a distinct trend toward valuing higher-order thinking or problem-solving skills, raising scientific hypotheses and questions by students as well as an increasing advocacy of epistemic knowledge for deepening learning in science. These radical shifts reflect a sensibility that what has served science education well in the past might not engender the kinds of critical scientific literacy (and forms of student agency) that these regions might need in the next 20 or 30 years. Readers will immediately see what we mean when they browse the country chapters; almost every system anticipates or has already welcomed revisions to their curricula and how elementary science ought to be organized. These accounts of the different trajectories of elementary science taken by these six regions will act like mini case studies to inform, inspire, and warn.

As well, affective aspects of science education that sustains interest in (lifelong) learning of science have been recognized as an important outcome of an education in science in a number of regions. This is an especially confounding topic for Korean (and Taiwanese) policymakers as Korean 4th graders' confidence and interest in science were very low compared to other OECD countries despite being a high-performing nation in the international tests. Likewise, Japanese children are now taught in school that it is necessary to have a "zest for life" that resonates well with allied efforts to make science fun for learners in Singapore and Hong Kong. Many authors in the edited book also pay tribute to the significance of informal science learning environments as well as science competitions in building motivation and interest. These nonschool sites where science can be part of a child's lifeworld and identity are hence increasingly attractive for primary science educators in East Asia.

Curriculum integration that includes science, technology, engineering, and mathematics (STEM) is a popular buzzword (e.g., in Hong Kong, China, Korea), although researchers might disagree with what passes for STEM in many classrooms around the world that is sometimes too truncated to be able to be truly interdisciplinary in nature. We thus think it is best to leave it to our readers' discernment as they proceed into the specific case descriptions. Indeed, when chapter authors here describe some of the challenges as well as accounts of failed projects, we are in an enviable position to benefit from the missteps or policy failures of others barring contextual differences (c.f. Zembal-Saul, this volume). It is very likely that each regional situation is far more complex and nuanced than what has been described here as much as we celebrate this book as a promising start in trying to get to grips with primary science education in East Asia.

To a large degree, the classroom assessment system has yet to keep up with changes in subject matter content and pedagogy; testing across the region remains largely conservative (mostly paper and pencil, less performance assessments), pervasive, and often overly consequential for future success. For example, Singapore has a high-stake examination at the end of Grade 6 that includes primary science besides testing in English, mother tongue, and mathematics. However, a deliberate recalibration of assessment that favors formative methods has been reported (e.g., Taiwan, Singapore, China) that bodes well for improving meaningful learning and learner autonomy. Korea has simply abolished the national assessment of primary children in 2013, and some provinces have instituted performance assessment tasks for greater authenticity. Japan too has been progressive in that national tests at elementary grades serve to evaluate the extent of learning in school/districts rather than being a proxy for measuring individual achievement. The national Taiwan Assessment of Student Achievement (TASA) for Grade 4 and 6 students functions in a similar manner, and the yearly results are made public.

Teacher training in East Asia is certainly not left to chance by governments, although larger systems (e.g., China) have multiple pathways for entry into teaching that might not be equally rigorous in their standards compared to smaller regions such as Singapore that has just a single institution for teacher education. It is also a mistaken assumption that becoming a primary science teacher is easy; Korea, for instance, mandates a tough and challenging testing system for would-be candidates that ensures only the most deserving qualify! Other regions have well-known high benchmarks for entry as described by various country authors; Hong Kong accepts candidates within the top 18% of each university cohort, while in Singapore, this figure hovers around 30%. Still, Korea and Singapore have explored the deployment of specialist teachers of science to mitigate the oft heard accusation of poor content mastery among primary teachers. In-service or professional development (PD) programs are an integral component of a teacher's life in East Asia; Taiwan, for instance, shared details of an accreditation program to recognize expertise in teaching, while teachers in Hong Kong and Singapore are entitled to 150 and 100 h of PD, respectively, per year. Knowing that up to 40% of elementary school teachers in Taiwan possess a master's degree is a testimony to the push toward gaining professional certification in that island.

Various science programs, reforms, and initiatives have been adopted by elementary science teachers in East Asia ranging from inquiry projects (e.g., Hong Kong, Korea, China, Japan), integrated studies (Japan) to the extensive use of ICT/makerspaces (e.g., Taiwan) and the use of informal science learning environments (e.g., Hong Kong)—the list is very extensive. What is happening with respect to the latter in China would not be unfamiliar to teachers in the Western world for Chinese teachers have been enthusiastically organizing summer science camps, robotics classes, and various outdoor activities to popularize science for young people. Of note, the use of science media and reading activities in Taiwan has taken prominence in efforts to leverage on the affordances of informal science learning.

That teachers in East Asia have been actively trying out different pedagogies is one obvious conclusion just as much as it reflects a strong culture of collaborative learning and sharing. Wednesday afternoons would find all Taiwanese elementary school teachers engaged in a full afternoon of PD, for example. Japan is exemplary as teachers there have been known to publish books to disseminate their successful teaching innovations and resources thereby spreading pedagogical content knowledge throughout the system. One method of pedagogical improvement research originating from Japan—lesson study—is in vogue and can now be found adapted for classrooms throughout East Asia. It is no exaggeration to state that teacher learning communities of all sorts are to be found here though termed differently. Yet, in larger Asian states, the rhetoric of reforms and equality of learning opportunities does not quite filter down to rural districts that experience a barebone version of science instruction as compared to what their urban counterparts routinely enjoy. As this problem is particularly acute in China, the state has publicly vowed to redouble its efforts to minimize the disparities within the urban-rural gap (see Normile 2017).

As editors, we felt justified to adopt a light editorial touch that allows the voices of the authors to be heard even though to the ears of native English speakers, the norms of fluency, elegance of expression, and rhythms of speech might have been sacrificed at many points. Not that we could have completely addressed all these concerns ourselves as we too are not native speakers of the language. We encourage readers therefore to look beyond the struggles of academic writing using a second or even a third language and to instead appreciate what has been achieved—a one-stop reference for primary science educational systems in East Asia.

1.2.2 Part II: The Expert Commentaries

Again, there is no particular order to reading as we had requested all the commentators to take a comprehensive view of the six regional chapters and then to offer their informed, critical thoughts. But to really do justice to as well as profit from Part II, it is advised to go through every one of these chapters. In toto, they are *the* editorial voice of this book! We did not influence or persuade them in any way, and what you see here represents their accumulated wisdom of theory and practice. Assembling such a spectacular lineup of commentators (and country authors) was no mean feat; they span all time zones in location and are the kinds of perceptive, honest, and clever people that one wished one had had mentors.

We have already mentioned the worldwide phenomenon of policy borrowing among Ministries of Education that increasingly practice governmentality (e.g., teacher accountability, high-stake testing, intensive training of teachers, etc.) at many levels. Economic considerations and societal improvement needs are additional though not exhaustive warrants often heard in educational systems undergoing reform. We are therefore pleased that research by Sinnema and Aitkin (2013) can offer us a kind of advance organizer as we approach these wide-ranging commentaries for they have identified major trends in curriculum reforms over the last decade. Similar to what has occurred in East Asia, curriculum goals among seven English-speaking countries have been reported to be (1) used as a lever for improvement, (2) used to serve equity issues, (3) future-focused, and (4) strengthened for coherence within and between levels. These have resulted in emphases on student competencies such as twenty-first-century skills that integrate knowledge, values, and attitudes with taking action as an outcome. Moreover, governments have made recommendations about how to teach subject matter that were frequently linked to outcome-based curriculum statements even as the former (contradictorily) sought to promote student agency and reduce prescriptions for teaching (Sinnema and Aitkin 2013). As Table 1.1 shows, these are some of the shared key themes in both general and primary science education in East Asia that our experts have chosen to highlight or review.

Now that Table 1.1 has provided a "rough guide" of the terrain, we can share in a little more detail the nature of their critique in this section. In their order of appearance, Adam Lefstein as our first critical friend starts off his provocative chapter by recalling Postman's (1995) *The End of Education* and how science education in each of these six regions might benefit when educators here discern the ends (a polysemous word!) of (science) education. Such metaphysical questions, although far from the pressing concerns of the average policymaker, are in fact the very goals that can inspire the learning of science for its own sake, for cultivating genuine scientific literacy and not just as a means to what Adam regards as a rather insipid end point: individual and national economic prosperity.

Terry Russell has tried to cover much ground, but has chosen to underscore an important question shared by many contemporaries in the field, "When and how do we enable pupils to think like scientists when faced with competing knowledge claims?" He takes us on a wide-ranging tour exploring what really matters at the heart of an education in primary science, namely, ideas about epistemic knowledge and scientific practices, promoting the nature of science and argumentation coupled

Table 1.1 Themes in general and primary science education in East Asia from the six expert commentaries

	Commentators					
Salient themes in general and primary science education in East Asia	Lefstein	Russell	Ramnarain	Symaco and Daniel	Zembal-Saul	McEneaney
Governmentality including teacher training, PD, accountability, testing	+	+	+	+	+	+
Curriculum lever for improvement		+		+		+
Serving equity issues, e.g., gender, age, disadvantaged	+	+			+	
Future-focused, e.g., 21 CC competencies		+		+	+	+
Questions of pedagogy and assessment in the curriculum		+	+	+	+	+
Student agency including activism for social and environmental	+	+		+		
Aims and goals of science education	+	+	+	+	+	+
Scientific literacy, nature of science, and higher-order or epistemic knowledge	+	+	+	+	+	+

with dialogic classroom environments. Terry also furnishes us with a repertoire of thoughtful ideas about emergent science, which is science for early years that East-Asian systems can adopt with profit.

Hailing from South Africa, Umesh Ramnarain offers food for thought regarding effective curriculum implementation and the related issue of teacher preparedness. These will together assist in delivering sound science lessons by instructors as well as give concrete expression to difficult reform efforts more broadly. He lauds the move by all East-Asian regions from a more content-based curriculum toward an inquiry-based one, although teacher readiness and working familiarity with inquiry teaching are truly nagging open questions during implementation. Readers will appreciate his scholarship as Umesh deeply engages with the relevant literature in his commentary.

Lorraine Pe Symaco and Esther Daniel take a different tack with in-depth discussions on common but significant themes that they have observed in the country chapters regarding science curricula, pedagogy, teacher education, and recent science education reforms. They also point to the inescapable importance of values, attitudes, and skills in the subject and, in particular, learning about environmental sustainability. Finally, Lorraine and Esther pose two uncomfortable questions as they close their multifaceted commentary: "How can primary science in these six regions leap to the next level? What does their future hold in terms of science education?"

An award-winning science educator, Carla Zembal-Saul, writes from her manifold experiences associated with supporting the development of both preservice and in-service elementary science teachers around the world. Explaining that East-Asian states were on the whole supportive of and encouraging of teachers' learning needs in science, coherence in teacher education was a tenuous condition that could be easily destroyed when schools, communities, districts, and other levels were misaligned in their goals/practices. Carla also does us a very useful service when she makes comparisons between the visions expressed in *A Framework for K–12 Science Education* and Next Generation Science Standards in the United States with aspects of science education policies and practices from East Asia, such as high-stake testing and student attitudes/interest in science.

For Elizabeth (Betsy) McEneaney, her focus on the broader institutional environments influencing what is taught and learnt in East-Asian science education derives from her vast expertise in sociological theory. By attributing a stable institutional order among these six East-Asian states, she explains with clarity and force how their curricula can use appropriate Internet and ICT resources (i.e., the technical environment) to augment scientific inquiry in schools while cultivating ties with disciplinary knowledge, language, and ideas. To Betsy, technology can be a disruptive game changer for the good! Looking at science education from these new angles will challenge us to revisit, if not radically alter, what we have held dear in science education.

References

Aldcroft, D. (2000). Education and development: The experience of the four little tigers. In A. J. H. Latham (Ed.), *Asia-Pacific dynamism, 1550–2000* (pp. 169–184). London: Routledge.
Allen, R. C. (2017). Lesson from history for the future of work. *Nature, 550*, 321–324. https://doi.org/10.1038/550321a.
Aloisi, C., & Tymms, P. (2017). PISA trends, social changes, and education reforms. *Educational Research and Evaluation, 23*, 180–220.
Ashton, D., Green, F., James, D., & Sung, J. (2000). *Education and training for development: The political economy of skill formation in East Asian newly industrialised economies*. London: Routledge.
Asian Development Bank [ADB]. (2015). *Key indicators for Asia and the Pacific: 2015 46th edition*. Manila: Asian Development Bank.
Atkin, M. J., & Black, P. (2003). *Inside science education reform: A history of curricular and policy change*. New York: Teachers College Press.
Benavot, A. (1992). Curricular content, educational expansion, and economic growth. *Comparative Education Review, 56*, 85–100.
Benavot, A. (2002). Education for learning to live together: A critical analysis of comparative research. *Prospects, XXXII*, 51–73.
Benavot, A., & Kamens, D. (1989). *The curricular content of primary education in developing countries*. Washington, DC: The World Bank.
Caillods, F., Göttelmann-Duret, G., & Lewin, K. (1997). *Science education and development: Planning and policy issues at secondary level*. Paris: UNESCO.
Cheng, M. H. M., & Wan, Z. H. (2015). Unpacking the paradox of Chinese science learners: Insights from research into Asian Chinese school students' attitudes towards learning science, science learning strategies, and scientific epistemological views. *Studies in Science Education, 52*, 29–62.
Chiu, M.-H. (Ed.). (2016). *Science education research and practices in Taiwan*. Dordrecht: Springer.
Chiu, M.-H., Guo, C.-J., & Treagust, D. (Eds.) (2007). Assessing students' conceptual understanding in science: An introduction about a national project in Taiwan. [Special issue]. *International Journal of Science Education, 29*, 379–390.
Darling-Hammond, L., Burns, D., Campbell, C., Goodwin, A. L., Hammerness, K., Low, E.-L., McIntyre, A., Sato, M., & Zeichner, K. (2017). *Empowered educators: How high-performing systems shape teaching quality around the world*. San Francisco: Jossey Bass.
Drori, G. (1993). The relationship between science, technology and the economy in lesser developed countries. *Social Studies of Science, 23*, 201–215.
Drori, G. (2000). Science education and economic development: Trends, relationships, and research agenda. *Studies in Science Education, 35*, 27–58.
Elyon, B.-S. (2014). Curriculum development. In R. Gunstone (Ed.), *Encyclopaedia of science education*. Available http://www.springerreference.com/docs/html/chapterdbid/303000.html
Forrestier, K., & Crossley, M. (2015). International education policy transfer – Borrowing both ways: The Hong Kong and England experience. *Compare, 45*, 664–685.
Harlen, W. (1994). Primary science. In T. Husén & T. N. Postlethwaite (Eds.), *The international encyclopedia of education* (pp. 5328–5335). Oxford: Pergamon.
Harlen, W. (2014a). Primary/elementary school science curriculum. In R. Gunstone (Ed.), *Encyclopaedia of science education*. Available http://www.springerreference.com/docs/html/chapterdbid/303020.html
Harlen, W. (2014b). Primary/elementary school science curriculum projects. In R. Gunstone (Ed.), *Encyclopaedia of science education*. Available http://www.springerreference.com/docs/html/chapterdbid/303021.html

Hollins, M., & Reiss, M. J. (2016). A review of the school science curricula in eleven high achieving jurisdictions. *The Curriculum Journal, 27*, 80–94.
Jenkins, E. W. (2015). Children and the teaching and learning of science: A historical perspective. In K. Schultheis & A. Pfrang (Eds.), *Children's perspective on school teaching and learning: Studies of the educational experience of children* (pp. 143–161). Zurich: LIT Verlag.
Jerrim, J. (2015). Why do East Asian children perform so well in PISA? An investigation of Western-born children of East Asian descent. *Oxford Review of Education, 41*, 310–333.
Khine, M. S. (Ed.). (2015). *Science education in East Asia: Pedagogical innovations and research-informed practices*. Dordrecht: Springer.
Kim, Y., Chu, H.-Y., & Lim, G. (2015). Science curriculum changes and STEM education in East Asia. In M. S. Khine (Ed.), *Science education in East Asia: Pedagogical innovations and research-informed practices* (pp. 149–226). Dordrecht: Springer.
King, K. (1989). Donor aid to science and technology education: A state of the art review. *Studies in Science Education, 17*, 99–122.
Lau, K.-C., & Lam, Y.-P. T. (2017). Instructional practices and science performance of 10 top-performing regions in PISA 2015. *International Journal of Science Education, 39*, 2092–2149.
Lee, Y.-J. (2008). Science education in and from Asia. *Cultural Studies of Science Education, 1–4*(OpEd), 3.
Lee, Y.-J. (Ed.). (2010). *The world of science education: Handbook of research in Asia*. Rotterdam: Sense Publishers.
Lee, Y.-J., Kim, M., & Yoon, H.-G. (2015). The intellectual demands of the intended primary science curriculum in Korea and Singapore: An analysis based on revised Bloom's taxonomy. *International Journal of Science Education, 37*, 2193–2213.
Lee, Y.-J., Kim, M., Jin, Q., Yoon, H.-G., & Matsubara, K. (2017). *East-Asian primary science curricula: An overview using revised Bloom's Taxonomy*. Dordrecht: Springer.
Liang, L. L., Liu, X., & Fulmer, G. (Eds.). (2017). *Chinese science education in the 21st century: Policy, practice, and research*. Dordrecht: Springer.
Lin, H.-S., Gilbert, J. K., & Lien, C.-J. (Eds.). (2016). *Science education research and practice in East Asia: Trends and perspectives*. Taipei: Higher Education Publishing.
Liu, X., Liang, L. L., & Liu, E. (Eds.) (2012). Science education research in China: Challenges and promises [Special issue]. *International Journal of Science Education, 34*, 1961–1970.
Loomis, S., Rodriguez, J., & Tillman, R. (2008). Developing into similarity: Global teacher education in the twenty-first century. *European Journal of Teacher Education, 31*(3), 233–245.
Manzon, M. (2014). Comparing places. In M. Bray, B. Adamson, & M. Mason (Eds.), *Comparative education research: Approaches and methods* (pp. 97–137). Hong Kong: Springer & Comparative Education Research Centre, The University of Hong Kong.
McEneaney, E. H. (2003). Elements of a contemporary primary school science. In G. S. Drori, J. W. Meyer, F. O. Ramirez, & E. Schofer (Eds.), *Science in the modern world polity: Institutionalization and globalization* (pp. 136–154). Stanford: Stanford University Press.
Normile, D. (2017). *One in three Chinese children faces an education apocalypse: An ambitious experiment hopes to save them*. Retrieved from http://www.sciencemag.org/news/2017/09/one-three-chinese-children-faces-education-apocalypse-ambitious-experiment-hopes-save
Pereyra, M. A., Kotthoff, H., & Cowen, R. (Eds.). (2011). *PISA under examination: Changing knowledge, changing tests, and changing schools*. Rotterdam: Sense Publishers.
Postman, N. (1995). *The end of education: Redefining the value of school*. New York: Knopf.
Rautalin, M., Alasuutari, P., & Vento, E. (2018). Globalisation of education policies: Does PISA have an effect? *Journal of Education Policy*, https://doi.org/10.1080/02680939.2018.1462890.
Sachs, J. (2005). *The end of poverty: Economic possibilities for our time*. New York: Penguin.
Sinnema, C., & Aitkin, G. (2013). Emerging international trends in curriculum. In M. Priestly & G. Biesta (Eds.), *Reinventing the curriculum: New trends in curriculum policy and practice* (pp. 141–163). London: Bloomsbury Academic.

Takayama, K. (2017). Imagining East Asian education otherwise: Neither caricature, not scandalization. *Asia Pacific Journal of Education, 37*, 262–274.

Wilkinson, G. (2013). McSchools for McWorld? Mediating global pressures with a McDonaldizing education policy response. In G. Ritzer (Ed.), *McDonaldization: The reader* (3rd ed., pp. 149–157). Thousand Oaks: Sage.

You, Y., & Morris, P. (2016). Imagining school autonomy in high-performing education systems: East Asia as a source of policy referencing in England. *Comparative Education, 46*, 882–905.

Yuen, S. (2017). Customising 'Singapore maths' for use in schools abroad. *The Straits Times Online*, 13 June 2017. Retrieved from http://www.straitstimes.com/singapore/education/customising-spore-maths-for-use-in-schools-abroad?login=true

Part I
The Regional Chapters

Chapter 2
Primary Science Education in Hong Kong

Wing Mui Winnie So, Zhi Hong Wan, and Yu Chen

2.1 Introduction

The recent collapse of a green roof shocked every citizen in Hong Kong as green roofs have been considered as providing great environmental benefits and have been installed in many local primary schools over the past 10 years (Hui 2006, 2011). These green roofs are commonly used as sites to involve pupils in urban farming; science, environmental, and community education have been implemented without much awareness of the disadvantages or safety risks of the green roof. Because of this incident, "school green roof safety" is now considered to be of great importance, regardless of its greenery effect. This is thus a pertinent opportunity for us to rethink issues that appear to have a positive effect on us but, in most cases, pose potential hazards and sometimes dreadful consequences as well. Thus, our children need to be equipped with the necessary scientific knowledge and skills starting from primary education to assess everyday and social issues, to have good reasoning for making sound judgment, and to voice their opinions in an informed and constructive manner.

Primary science is a core subject in many national school systems. Why is primary science important? Harlen (2008) in her writing about science as a key component of the primary curriculum stated that the 1950s were a watershed in school science education in western countries due to the demand for scientists and technologists as a result of the speed of scientific development. She reported the two points made by Nathan Isaacs in 1962 regarding the preeminence of "processes" of science that science has brought and is still developing scientific knowledge and the hope that science education could somehow affect people's reasoning in everyday life.

In Hong Kong, primary science was introduced as Nature Study in the 1970s, as primary science in the 1980s, and as the science education key learning area in an

W. M. W. So (✉) · Z. H. Wan · Y. Chen
The Education University of Hong Kong, Hong Kong, Hong Kong
e-mail: wiso@eduhk.hk

Y.-J. Lee, J. Tan (eds.), *Primary Science Education in East Asia*, Contemporary Trends and Issues in Science Education 47,
https://doi.org/10.1007/978-3-319-97167-4_2

integrated subject called General Studies in the 1990s (So 1997; Chan and Lui 1998). What was the impact of these curriculum changes in primary science on pupils' science learning? There was no detailed analysis of this until 2008 when So (2008) analyzed the TIMSS data of the past 20 years from the 1980s to 2003 and found that the international ranking of Hong Kong Primary 4 pupils had increased significantly during these 20 years, reflecting the improvement in pupils' science achievement and the possible effects of the reform of the science curriculum on teaching and learning. However, which components in the curriculum and teaching and learning led to such improvement has not been well studied. Moreover, according to the TIMSS reports, Hong Kong 4th graders have been gradually improving in their science achievement from 1995 till 2007, with a slight decline in 2011[1], but a rise of students' ranking in science from ninth to fifth from the most recent TIMSS 2015[2] report. These interesting findings have stimulated us to take a closer look at the development and state of primary science education in Hong Kong.

This chapter starts with an overview of basic education of Hong Kong as the background to understanding primary science education. This is followed by an account of the history of primary science education to help understand the development and changes that have taken place over 50 years. Next is an outline of the current primary science education in the integrated curriculum of General Studies, focusing on the curriculum organization, emphasis, and suggested pedagogy. Pupils' informal learning opportunities in science events as extracurricular activities will provide some hints of the impact of science education on pupils' science process and science thinking. The status of teacher education in science provides some background to the competence of teachers in science teaching. Lastly, there is a review of the research or studies conducted on the implemented curriculum reflecting the concern of science educators regarding primary pupils' science learning.

2.2 Overview of Basic Education in Hong Kong

To meet the challenges of the twenty-first century, Hong Kong is in need of versatile talents, and school education is a deciding factor in the success of nurturing such talents. Hence, based on the following recommendations made by the Education Commission (EC) and Curriculum Development Council (CDC): EC's *Learning for Life, Learning through Life – Reform Proposals for the Education System in Hong Kong (2000)*[3]; CDC's report *Learning to Learn – The Way Forward in Curriculum Development (2001)*[4], *the Basic Education Curriculum Guide – Building on*

[1] http://www.iea.nl/publications.html

[2] http://www.info.gov.hk/gia/general/201611/29/P2016112900718.htm

[3] http://www.e-c.edu.hk/eng/reform/annex/Edu-reform-eng.pdf

[4] http://www.edb.gov.hk/attachment/en/curriculum-development/cs-curriculum-doc-report/learn-learn-1/overview-e.pdf

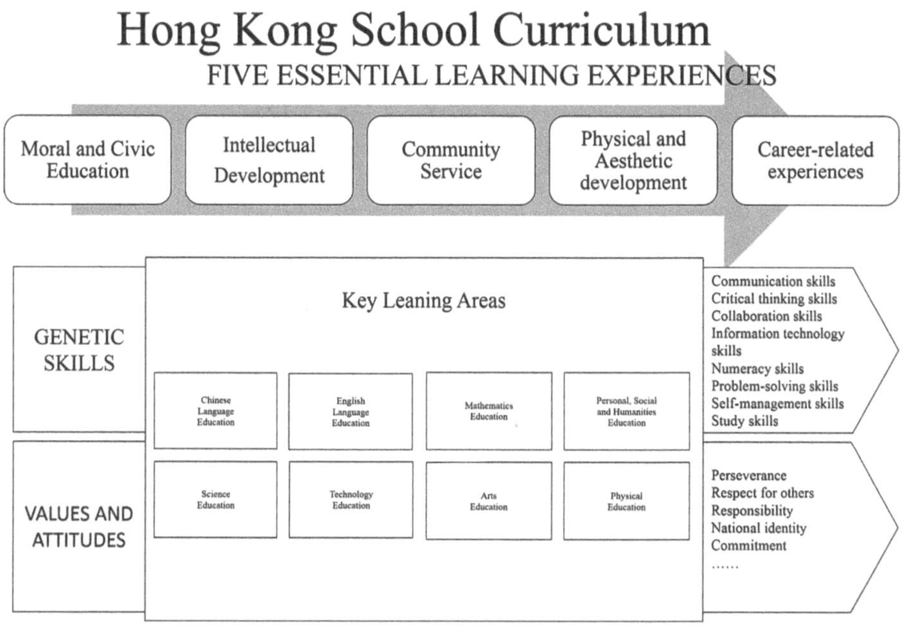

Fig. 2.1 The curriculum framework of the Hong Kong school curriculum. (Adapted from CDC 2014, Fig. 1.3)

Strengths (Primary 1 - Secondary 3) (2002)[5], the direction of the curriculum reform launched in the same year is to provide comprehensive and balanced learning experiences for pupils. In addition to acquiring knowledge in class, pupils are expected to develop "learning to learn" capabilities as well as "positive values and attitudes" for achieving the educational aims of "whole-person development" and "lifelong learning" (CDC 2002a).

Figure 2.1 illustrates the curriculum framework[6] of the Hong Kong school curriculum which is comprised of three interconnected components: (1) knowledge in key learning areas, (2) generic skills, and (3) values and attitudes. This is claimed to be an open curriculum framework which allows schools to organize and offer learning experiences at different paces, adjust the breadth and depth of learning content, and flexibly adopt a range of learning strategies and modes to maximize learning and teaching effectiveness.

A series of eight key learning area (KLA) curriculum guides (primary 1 to secondary 3) have been developed by the Curriculum Development Council (CDC) to support the curriculum reform. Although the conventional subjects of English, Chinese, and Mathematics are still used in schools, the concept of key learning areas is employed in the curriculum. There are eight key learning areas (KLAs): Chinese language education; English language education; mathematics education;

[5] http://cd1.edb.hkedcity.net/cd/EN/Content_2909/BE_Eng.pdf

[6] Basic Education Curriculum Guide https://cd.edb.gov.hk/becg/english/chapter1.html

personal, social, and humanities education; science education; technology education; arts education; and physical education.

Building on the practical experiences and fruitful outcomes of the development of the primary school curriculum for almost a decade, schools are recommended to sustain school-based curriculum development by building on their existing strengths by referring to the updated *Basic Education Curriculum Guide – to Sustain, Deepen and Focus on Learning to Learn (Primary 1–6)*[7] released in 2014 and the updated curriculum guides of the related KLAs released in 2017.

The position of the KLA of science education[8] in the government document for school children from primary 1 (CDC 2002b) is that science is the study of phenomena and events around us through systematic observation and experimentation (p. 3). Science education cultivates pupils' curiosity about the world and enhances scientific thinking. Through the enquiry process, pupils will develop scientific knowledge and skills to help them evaluate the impact of scientific and technological development. This will prepare pupils to participate in public discourse in science-related issues and enable them to become lifelong learners in science and technology (Lee and Ng 2004).

The emphasis of science education is to enhance pupils' scientific thinking through progressive learning activities that involve planning, measuring, observing, analyzing data, designing and evaluating procedures, and examining evidence. Learning science will enable our pupils to lead a fulfilling and responsible life by encouraging them to learn independently, deal with new situations, reason critically, think creatively, make wise decisions, and solve problems (CDC 2002b).

Through science activities, pupils should develop an interest in science, and thus they will be motivated to become active learners in science. Pupils should also develop an understanding of science, technology, and society (STS). They should be able to make informed decisions based on evidence. Figure 2.2 shows the diagrammatic representation of the science education framework.

2.3 Historical Background of Primary Science Education in Hong Kong

The development of primary science education in Hong Kong has three important stages. In the 1970s, primary science was introduced as Nature Study and Rural Study, and then the Primary Science Curriculum replaced Nature Study in the 1980s. More recently, the science education key learning area in the integrated subject of General Studies covering primary science education was implemented.

[7] Updated *Basic Education Curriculum Guide – to Sustain, Deepen and Focus on Learning to Learn (Primary 1–6)* https://cd.edb.gov.hk/becg/english/intro.html

[8] KLA of Science http://www.edb.gov.hk/en/curriculum-development/kla/science-edu/index.html

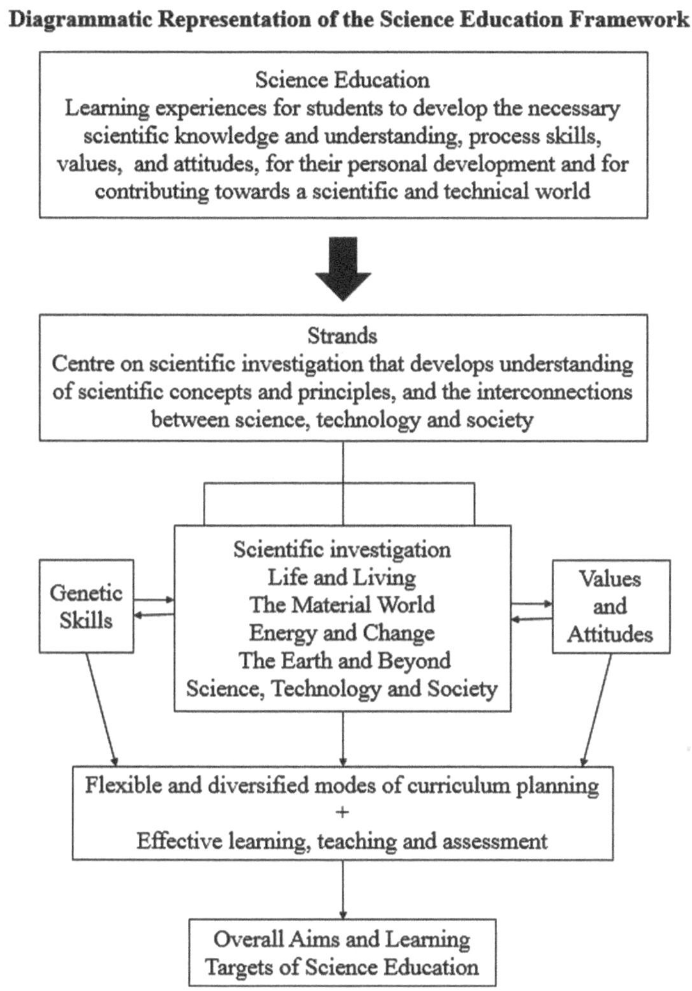

Fig. 2.2 Diagrammatic representation of the science education framework (Adapted from CDC 2002b, p. 19)

2.3.1 Nature Study and Rural Study in the 1960s–1970s

In 1967, the Nature Study curriculum for urban pupils and Rural Studies for rural pupils were introduced as the first central common-core primary science curricula in Hong Kong. The Nature Study curriculum (Education Department 1967) aimed to:

- Stimulate the child's curiosity about nature.
- Help the child know his/her expanding environment.
- Bring the child into contact with nature and develop his/her powers of observation.

- Foster a considerate attitude toward living things and cultivate understanding and appreciation of natural phenomena.
- Help the child live a fuller life by putting into practice his/her knowledge of nature.

It was recommended wherever possible that direct observation supported by objects, specimens, and cuttings from magazines and periodicals be used. A nature corner with nature table, shelf, or bench with updated exhibits and an aquarium should be available. During that time, children were encouraged to bring their own pets to school. Laboratory and special apparatus were not required, but a large table for demonstration purposes and simple apparatus made by teachers and pupils could be used. Moreover, it was suggested that Nature Study teachers cooperate with Geography teachers to avoid duplication of work.

The majority (two-thirds) of the curriculum content was biology, with topics related to physical science, earth science, and other areas of science and technology not introduced until primary 5 and 6. Although firsthand experience was emphasized in the curriculum, the "matchbox" schools in the 1970s with poor school facilities and the lack of science equipment did not allow much chance for primary school children to engage in firsthand learning experiences. In Lo's (2002) analysis of the curriculum, the ideological elements appear more as educational rhetoric rather than as a real task-oriented model of learning.

2.3.2 The Primary Science Curriculum in the 1980s

The Science Subject Committee (primary) of the Curriculum Development Committee (1981) held its first meeting in 1976 to design a Primary Science Curriculum to replace the existing Nature Study and Rural Studies curricula. The Committee, with membership including heads and teachers from government and nongovernment schools; representatives of universities, colleges of education, and other organizations; and officers of the advisory inspectorate and other divisions of the Education Department, took 3 years to complete the draft, and a large-scale trial teaching with 80 primary schools was conducted in 1980 to better understand the implementation feasibility of the curriculum and the possible obstacles encountered. The draft was further revised based on the report of the trial teaching, and the Primary Science Curriculum was issued in 1981.

The design of this curriculum was to widen children's sphere of learning experience through their exploration of the natural environment with the teacher's role to engage pupils in discovering scientific truth for themselves (CDC, 1981). Lo (2002) reported that, compared to its 1960s counterpart, the science curriculum developed in the 1980s began to acknowledge its role in promoting the personal development of children, although it still emphasized the cultivation of disciplinary knowledge and skills. Therefore, there was the emergence of a "child-centered" science curriculum in the 1980s. However, Holbrook (1990) observed that with limited teaching time for this subject and a lack of proper science teacher training, the teaching focus of primary science was neither experimental science nor science processes. However, it was discovered that there was an overlap of teaching content among the three curricula of primary science, health education, and social studies. The following are two examples of related/similar content under the same topics of "electricity" (Table 2.1) and "water" (Table 2.2) with content taught in the three subjects.

2.3.3 General Studies from 1996

Primary science as a single subject was implemented for nearly two decades in primary schools from the early 1980s until the introduction of an interdisciplinary subject, General Studies (GS), in 1996. Such change was in response to problem raised by the Hong Kong Education Commission (EC 1990, p. 13) that "the range

Table 2.1 Topics related to "electricity"

Class Level	Social studies	Primary science	Health education
P.1			Home safety
P.2		How to use dry cells?	
P.4	Knowing the facilities among us		Home safety
			Fire safety
	Electricity and fuels		Simple first aid
P.5		What is static electricity?	
		Can energy produce heat?	
		Can energy produce magnetic force?	
		What are the uses of electricity?	
		What are the precautions of using electricity?	
P.6	Contributions of technology to humans		
	Investigation of steam engines		
	Use of electricity		
	Influences of technology on human life		

Table 2.2 Topics related to "water"

Class Level	Social studies	Science	Health education
P.2		What are the forms of water?	
P.3			Importance of drinking water
P.4	Water supply	What are the characteristics of water?	Water supply
P.5		How are fog, dew, and frost formed?	Water and fiber

of discrete subjects offered (by the common core curriculum) is too wide resulting in fragmentation and compartmentalization of the curriculum." In order to reduce repetition in curriculum content and to give schools greater flexibility and curriculum "space" to let pupils learn more comprehensively and effectively, GS was introduced as a subject that integrated the learning elements of three former subjects: health education (CDC 1980a), social studies (CDC 1980b), and primary science (CDC 1994). Hong Kong is not the only place implemented an interdisciplinary curriculum with primary science learning. In the International Review of Curriculum and Assessment Framework Project (INCA) in 2000, 13 of the 18 countries participated in the review were having a compulsory subject of "science" or "science and technology," while the others were teaching science with social or environmental studies for primary years (O'Donnell and Michlethwaite 2000; O'Donnell 2002).

The aims of GS are to provide children with the basic knowledge related to the individual, family, society, science and technology, and the natural world in the four content areas of "Heathy Living," "Living Environment," "the Natural World," and "Science and Technology." The focus on the integrated nature of the GS curriculum aroused educators' attention, and there were several related studies conducted after the subject was implemented. One study measured teachers' receptivity of the subject, and the responses were found to be negative (Lee 1998). So (1998) also identified difficulties faced by primary school teachers during the teaching of science-related topics in GS. The main obstacles faced by teachers were identified by So et al. (1998a, b) and included insufficient subject knowledge, inadequate resources, lack of experience, and difficulties handling pupils' learning problems. Teachers' difficulties in teaching science would negatively impact the quality of GS implementation and subsequently hinder the development of students' scientific literacy.

In the second year of the curriculum implementation, there were studies which investigated teachers' views on the curriculum integration and their understanding of integrated subjects (So et al. 1999). They commented that "when compared with the spiral approach used in the three conventional subjects, the current arrangement of the General Studies syllabus was not more effective in terms of facilitating pupils' learning" (p.22). The one-off arrangement on most of the topics in GS was not structured for children to gain interests in some topics at lower primary levels and to gain better understanding of the related concepts at upper primary. This is in contrast to innovations in the USA and the UK which have taken children's progression in learning science into consideration in the curriculum design.

After 5 years of implementation, the perceptions and practices of GS teachers were evaluated by Lo and Cheng (2001) using a questionnaire, which made useful findings to facilitate further curriculum development. The results found that the majority of the GS teachers would be able to recognize the need of the curriculum integration, but the lack of adequate pedagogical knowledge related to the integrated curriculum would hinder their practices.

Research studies continued during the revision of the GS curriculum in 2002. A case study of a beginning teacher conducted by Cheng and Lo-Fu (2002) shed light on teachers' conceptions, the organization and structure of the GS curriculum, concepts and skills in thematic teaching, as well as teaching strategies. Based on the suggested inquiry approach in the learning and teaching of GS, So (2002) proposed a curriculum design from the viewpoint of the constructivist learning theory and indicated the need to link the GS curriculum with pupils' daily life.

A revised GS curriculum guide was developed in 2002 (CDC 2002c) and was implemented in 2004 in accordance with the curriculum reform in Hong Kong schools that began in 2001. The suggested curriculum development in GS is moving away from the content-focused approach and toward the learner-focused approach to enhance the inquiry and investigative skills of learners to help them construct knowledge, with an emphasis on the use of diverse resources and lifelong learning opportunities. The GS curriculum was further updated and enriched in 2011 and placed more emphasis on the preparation of students' development of a healthy lifestyle, a respect for self and others, a responsibility to environment, and an ability to use technologies (CDC 2011).

An updated curriculum guide was released in mid-2017[9] with reference to the updated *Basic Education Curriculum Guide – to Sustain, Deepen and Focus on Learning to Learn (Primary 1–6)* with a new emphasis on STEM education.

2.3.4 Primary Science Education in the Integrated Subject of General Studies

Currently, GS is an interdisciplinary subject comprising the KLAs of personal, social, and humanities education (PSHE), science education (SE), and technology education (TE), which provides pupils from P1 to P6 with opportunities to integrate knowledge, skills, values, and attitudes across the KLAs (Fig. 2.3). The GS curriculum provides learning experiences:

1. To help pupils have a better understanding of themselves and the world around them
2. To arouse pupils' interest in and development of their skills to enquire about science, technology, and society
3. To cultivate positive attitudes and values for healthy personal and social development (CDC 2011, p. iii)

Teachers act as facilitators of pupils' learning, are encouraged to adopt diversified learning and teaching strategies, and involve pupils in firsthand scientific investigations. The central GS curriculum, in the form of an open and flexible curriculum framework, sets out the learning targets and learning objectives for the development of pupils' subject knowledge, generic skills, positive values, and attitudes under the following six strands, rather than the previous four major content areas. Instead of teaching the isolated core elements from different KLAs, schools are encouraged to

[9] General Studies Curriculum Guide for Primary Schools (P1-P6)(2017)http://www.edb.gov.hk/en/curriculum-development/cross-kla-studies/gs-primary/curriculum-documents.html

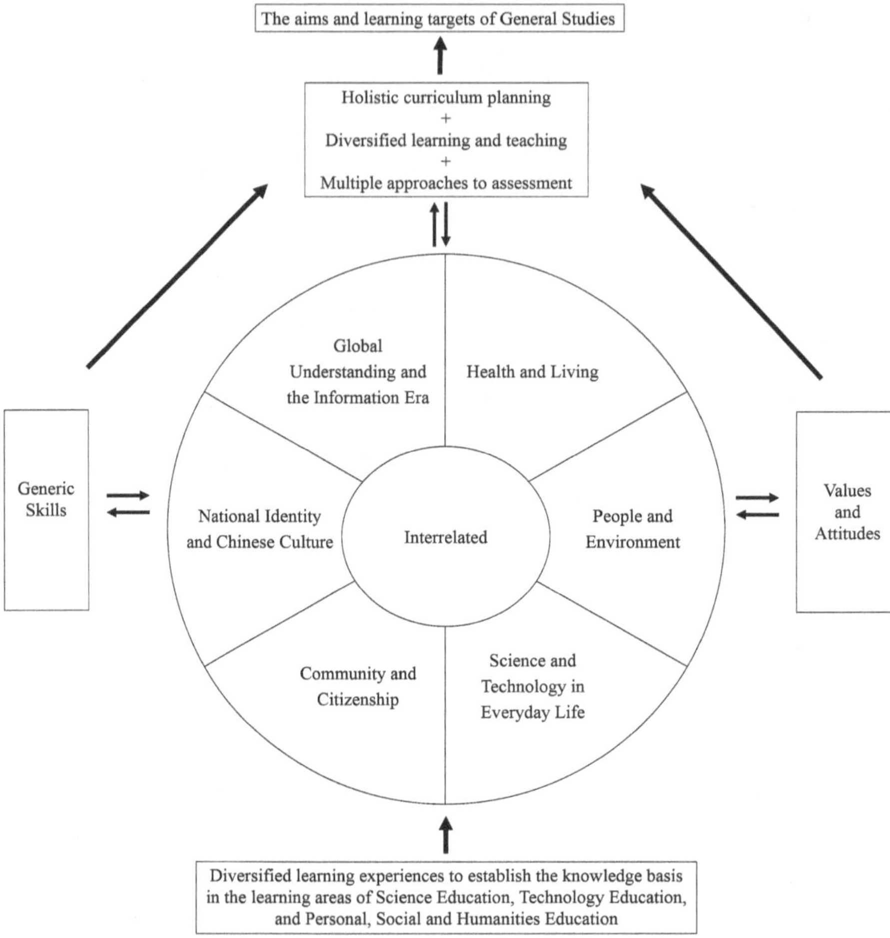

Fig. 2.3 The curriculum framework of GS (Adapted from CDC 2017, p.10)

explore the different themes integrating the core elements of the curriculum so as to meet the needs and interests of their pupils and update them in accordance with social change.

Taking the guiding principles into consideration, teachers are encouraged to adopt diversified learning and teaching strategies, of which enquiry-based learning is a commonly used approach. The major strategies of project learning and scientific investigation have been used with success in schools, both local and overseas.

Schools are advised to allocate around 12–15% of pupils' learning time to the learning of GS, with the majority of schools allocating four to five 35-min GS lessons in a week. It is expected that there will be a balanced coverage of the six strands, with pupils engaging in learning experience of all of the core elements suggested in the curriculum document. Yet, the time available for science learning is still very limited. For those schools which are enthusiastic with science learning, teachers usually arranged science investigation projects day/week/period once or twice in a year. Examples of these successful stories are shared in the session "Chief Executive's Award for Teaching Excellence in Science Education."

2.3.5 *Opportunities for Science Learning Outside the Classroom*

To address the call for more opportunities for pupils to engage in science inquiries, there are different types of science events and science fairs for primary pupils, mainly aged from 6 to 12, in Hong Kong. Most are organized by different public or private organizations, including the Hong Kong Science Museum, and can be divided into two major types, workshops and competitions, which are organized outside school hours.

Competitions require more complicated investigations of science, for example, from the stage of investigation design, result recording, and measurement through to data interpretation. Participants can strengthen their knowledge and skills integration ability and increase their creativity, collaboration, and problem-solving skills. In recent 2 years, there have been three major science competitions, with only one of them mainly for primary pupils, the "Primary STEM Project Exhibition" (PSPE), co-organized by several organizations including the Education University of Hong Kong, the Hong Kong Science Museum, the Education Bureau, the Hong Kong Education City, and the Hong Kong Association for Science and Mathematics Education. The other two competitions organized by individual organizations are for primary, secondary, and even kindergarten pupils. Table 2.3 provides a summary of these competitions with a brief description, the organization which runs them, the target year level, and remarks.

Workshops, on the other hand, usually involve simple and interesting hands-on activities or experiments by which participants learn particular science concepts or knowledge through the process of doing the hands-on activities and interactive discussion with participants and instructors. One such event is "InnoCarnival," organized by the Innovation and Technology Commission, which includes several workshops such as "Light Up! Paper circuit for makers," "Robot Fish – Horizontal style," "Salt painting," and "Colour My Chromatograph." Another big event, the HK SciFest 2017, was organized by the Science Museum. The "Vibrant Science Academy – summer science party" organized by the Hong Kong Science and Technology Parks Corporation; "Little Scientists Exploration Programs: powerful force" organized by the Hong Kong Science Museum; "Rock Classroom" organized by the Hong Kong Global Geopark, Agriculture, Fisheries, and Conservation Department; "Family Workshops – the final frontier discovering the earth" co-organized by the Science Museum, Education Bureau, and Hong Kong Education City; The "Moon Mission" organized by the Ho Koon Nature Education Cum

Table 2.3 Details of competitions organized for Hong Kong primary pupils

Name of event	Description	Organization	Target	Remarks
Primary STEM Project Exhibition (PSPE)	The "Primary STEM Project Exhibition" (PSPE) is the largest annual event featuring primary school science, technology, engineering, and mathematics (STEM) projects committed in nurturing talents in science exploration	Education University of Hong Kong	Primary 4–6	The PSPE has been held since 1997 and has been well received by the stakeholders in the education field. The project has become a platform for sharing science and STEM inquiry outcomes among primary pupils in Hong Kong and the Pearl Delta Regions of Mainland China, with over 100 schools and 1000 pupils participating each year
	https://www.hkedcity.net/pspe/	HK Science Museum		
		Education Bureau		
		HK Education City limited		
		HK Association for Science and Mathematics Education		
Hong Kong Youth Science and Technology Innovation Competition	The competition has categories including science fiction drawing, invention, research, science and technology practical activity, outstanding science and technology teacher, and outstanding science and technology school	Hong Kong New Generation Cultural Association	Kindergarten pupils, primary and secondary school pupils, teachers, lab technicians, and teaching assistants	Since its launch in 1997, this competition is now the largest and most popular science competition in Hong Kong with nearly 4000 applicants from over 400 schools participating in this annual event
	http://stic.newgen.org.hk/load.php?link_id=60286			
GreenMech Contest	The program includes stage design, structure design, material choices, and actual project making. All these procedures are completed in teams within a set time limit; in this sense, pupils get the chance to apply what they know and what they have learned from school that is relevant to the architecture of modular toys	The HK Federation of Youth Groups	Primary 4–6	To foster teenagers' creativity in mechanical design and scientific applications, the HK Federation of Youth Groups first introduced the "GreenMech Contest" from Taiwan in 2010 and has so far attracted more than 850 participants from 83 primary and secondary schools and education centers
			Secondary 1–3	
	http://ccst.hkfyg.org.hk/page.aspx?corpname=ccst&i=1130		Secondary 4–6	

Astronomical Centre; "Program of Toy Robot: dancing to music" organized by Smart Kiddo Education Ltd. in association with the Science Museum; "Science DIY: hands-on workshop of science gadgets" organized by the Polytechnic University with the Science Museum; and the "Young Astronaut training camp – foundation project" organized by Solar Tower Camp are all workshops catering to primary-level pupils. Evaluations of these workshops and activities were mostly conducted by the organizers and participating pupils usually reported interesting and useful learning experiences with them (So 2015).

A rich database of students' science inquiry projects has been developed with the annual event PSES; booklets each with over 20 outstanding students' project in previous 20 events have been uploaded in the website[10] to create a platform for sharing ideas of science and environmental learning. The great impact of this event to primary science education in Hong Kong was well received with a Grant Award in Knowledge Transfer[11] awarded by the Education University of Hong Kong in 2015.

2.4 Teacher Education in Science Education

As commented by Tucker's (2016) reporting Hong Kong in the crosshairs of the future[12], Hong Kong is having an excellent teacher corps. With the introduction of Bachelor of Education programs, the quality of those who go into teaching is improving as entering into education program is among the 18% getting into universities.

In order to equip teachers to be competent in teaching GS, there are initial teacher education programs, including the Bachelor of Education (BEd) for primary pupils and the Postgraduate Diploma of Education (PGDE) for degree holders who would like to pursue a career as a GS teacher, provided at the five universities, the Hong Kong University, the Chinese University of Hong Kong, the Education University of Hong Kong, Hong Kong Baptist University, and the Open University of Hong Kong, while the Education University of Hong Kong is the major provider. Generally, graduates with a BEd qualification or graduates holding an undergraduate degree plus a PGDE degree are qualified to become teachers in Hong Kong. Moreover, there are professional development (PD) programs for in-service teachers to update their pedagogical knowledge and skills with the different KLAs.

[10] Primary STEM Project Exhibition (PSPE) https://www.hkedcity.net/pspe/

[11] Primary STEM Project Exhibition (PSPE) https://www.youtube.com/watch?v=3Tk_vFPxAjM

[12] http://blogs.edweek.org/edweek/top_performers/2016/06/hong_kong_in_the_crosshairs_of_the_future.html

2.4.1 Initial Teacher Education: Bachelor of Education (Honors) (Primary) – General Studies Program

The Bachelor of Education (Honors) (Primary) – General Studies program gives the student teachers opportunities to acquire subject knowledge related to science education for teaching and learning of General Studies in primary classrooms. Among the 39 major credit points (cps) in the program, there are courses (worth 3 cps each) on "Environmental Studies"; "Healthy Living"; Living in the Information Age"; "Children's Science Learning"; "Science in the Contemporary World"; "the Natural World"; "Science, Technology, and Society"; "Interdisciplinary Concepts and Thinking"; and "Trends and Development in General Studies." Particularly, the student teachers have the opportunity to learn science in the course under the strand "Nature, Science & Technology" in the General Education Breadth courses.

In addition, there are 6 cps for Education Studies (with a major in Pedagogy) which touch on teaching methodologies. Moreover, the Honors Project, a 3-cp course, is a 1-year project usually conducted in the final year of the BEd program. Pupils can choose their topic from either their major or Education Studies to complete an individual research project. Since this Honors Project is supervised by an academic staff member, most students choose to work on their major.

However, a study by So et al. (1998a, b) showed that the majority of the General Studies teachers do not have a science background in their secondary education. The teachers who do not have a non-science background are likely to encounter more difficulties in teaching science-related topics than those with a science background, particularly in responding to students' questions, preparing teaching materials or tools, clarifying scientific concepts, focusing student attentions on learning, and organizing scientific investigations in the classroom.

2.4.2 Teacher Professional Development in Science Education KLA

In Hong Kong, there is a requirement that every teacher has to complete 150 h of professional development activities in a 3-year cycle. It was reported in the Advisory Committee on Teacher Education and Qualification (2009) that teachers mostly invest over 60% of the professional development hours required in "Teaching and Learning" activities. The teacher professional development opportunities for Hong Kong teachers are usually provided and funded by the Education Bureau in the form of tender quotations. However, teachers do not have the chance to try out the teaching pedagogy they have learned in these programs, which are usually short term and ad hoc. A study on the professional development of primary school teachers in teaching science (So and Cheng 2000) identified the professional development needs for improving science teaching, and the study puts forward some suggestions for directions to help frame teacher professional development programs.

The following are the tendered teacher development courses provided by the team of science educators from the Education University of Hong Kong. They are mostly one or two full-day workshops, based on the tendering requirements of the Education Bureau, and are designed to equip teachers with some basic scientific understanding and provide them with hands-on opportunities for designing some practical science activities:

- "In-service teacher development courses on science and technology in General Studies" (2008, 2009, 2010)
- "In-service teacher development course on Scientific Investigation in General Studies" (complete in 2011, 2012, 2013, 2014)
- In-service teacher development course – 1-day workshop on science and technology in General Studies (2015/16)
- In-service teacher development course on "1-day workshop on curriculum planning, pedagogies, and assessment of STEM education in General Studies (2016/2017)

Except for new education innovations, the Education Bureau does not invest significant funding in other teacher professional development such as university-school partnerships (Cheng and So 2011) even though extensive research has been carried out on their merits in terms of enhancing professional dialogue.

2.4.3 *Chief Executive's Award for Teaching Excellence (ATE) in Science Education*

Teachers can also learn from each other through some good practices in science education from awardees for teaching excellence.

Funded by the Quality Education Fund and organized by the Education Bureau, the CEATE was designed to recognize model teachers who excel in their teaching practice, to raise the status and public respect for teachers, and to allow good teaching practices to be shared with larger audiences[13]. Originating in 2003[14], CEATE recognized excellent in-service teachers in different domains in each round. Since 2003, the awardees from the academic years 2012–2013 and 2005–2006 were from science education as one of the eight key learning areas.

All awardees (including the Award for Teaching Excellence and the Certificate of Merit) are given trophies, a cash award, a dissemination grant, and continuing professional development scholarship[15]. All awardees join the CEATE Teachers Association and share their excellent teaching practices with teachers in other schools.

---

[13] P.6, Chief Executive's Award for Teaching Excellence Compendium 2005–2006

[14] http://www.ate.gov.hk/english/ate_previous1.html

[15] P.6–7, Chief Executive's Award for Teaching Excellence Compendium 2012–2013

Table 2.4 Selection criteria for CEATE (reference)

Area		Skills/details
Professional competence	Curriculum	Curriculum design and organization
		Curriculum management
	Teaching	Strategies and skills
		Professional knowledge and attitude
	Performance assessment	Assessment planning and use of information
Pupil development		Attitude
		Knowledge and skills
Professionalism and commitment to the community		Contribution to the profession and the community
School development		Support to school development

The selection criteria are based on the following four domains: professional competence, pupil development, professionalism and commitment to the community, and school development. Details of each domain are summarized in Table 2.4.

In 2005–2006, among the 11 teams/individuals awarded the ATE in the science education key learning area, 3 were from primary schools. For the Certificate of Merit, four teams among the six teams/individuals were awarded. The highlights of the respective awardees are as follows[16]:

- Seek truth, goodness, and beauty via environmental education.
- Recognize motivation, and cultivate a spirit of scientific investigation.
- Have fun in science, and stimulate the learning interest of the pupils.
- Promote lively pedagogy via collaborative thinking.
- Apply knowledge to daily life through inquiry-based learning.

In 2012–2013 (science education key learning area)[17], one primary teacher among the four teams/individuals was awarded the ATE award, while two primary school teams among the five teams were awarded the Certificate of Merit. The highlights of the awardees are as follows:

- Cultivate curiosity from a young age via teacher-led scientific inquiry.
- Have no limit to scientific inquiry via online classrooms.
- Grasp the characteristics of pupils, and design classes that are fun.

Cheng and So (2009), by interviewing the principals, panel heads, and teachers of two schools awarded the CEATE, reported the strategies and difficulties of providing quality primary science education through an integrated curriculum. It was found that the schools employed a team preparation approach and continuous professional development to raise teachers' confidence in teaching science. In order to increase the status of science in the primary curriculum, there was a need for the

[16] P.8, Chief Executive's Award for Teaching Excellence Compendium 2005–2006

[17] P.9, Chief Executive's Award for Teaching Excellence Compendium 2012–2013

school principals to value science teaching and learning. To overcome the insufficiency of curriculum time and recourses, the primary teachers would be encouraged to use extracurricular activities. The views of the importance and priority these teachers gave to science in light of the competing demands from various education reform initiatives were reported. The strategies employed for enhancing pupils' performance and understandings in science are reflected, providing good references for other schools and teachers in Hong Kong.

2.5 Research in Primary Science Education

In addition to the research into the challenges and opportunities of implementing science education in General Studies and the TIMSS comparative study of science learning, a good number of works have been undertaken to explore science teaching and learning in Hong Kong primary schools. These works broadly fall into five areas, (1) pupils' understanding of science concepts, (2) scientific inquiry, (3) science teachers and their development, (4) information technology (IT) in science education, and (5) environmental education. It should be noted that a number of the papers included in the 4th area, i.e., information technology in science education, are also related to scientific inquiry and teacher development. However, considering IT as a new and flourishing area, these works are integrated with other works on IT in science education to reflect the trend of the development in the research landscape.

2.5.1 Pupils' Understanding of Science Concepts

Focusing on pupils' understanding of air resistance, Lee and Kwok (2009) explored Grade 6 pupils' (aged 11–12) learning when conducting an egg-dropping race. The finding revealed that only a relatively small proportion of pupils achieved a better understanding of the concept after the race. They still had misconceptions about how air resistance is affected by the mass and size of falling objects. It is suggested that the concept should be made more visible and be connected to pupils' daily life experiences. Lin (2015) investigated 102 fifth graders' understanding about salt and electricity in computer-supported knowledge-building environments through case studies and found that such learning environments would contribute to pupils' conceptual change. However, some alternative conceptions existed. For instance, some pupils were under misconceptions that salt can conduct electricity or salt water cannot conduct electricity. Suggestion is that science teachers could consider to create epistemic-enriched learning environment to promote pupils' scientific epistemology and subsequently their conceptual understanding.

2.5.2 Science Inquiry

Inquiry-based learning is explicitly emphasized in the General Studies for primary schools' curriculum guide (Primary 1–Primary 6) as a student-centered approach which helps students to integrate generic skills, knowledge, and values. In the inquiry process, "students are active constructors of knowledge and the teacher is a facilitator of learning. ... students have to raise questions, find their own answers and look for the necessary information. They are engaged in identifying problems, collecting information and solving the problems they encounter" (pp. 75–76).

Science inquiry has begun to be an important area of research in primary science education in Hong Kong. The initial efforts of research on scientific inquiry focused on generating a theoretical model to conceptualize the process of scientific inquiry. So (2003) analyzed 24 written records of scientific inquiry conducted by groups of Hong Kong primary school children which were submitted to the final of a primary science project event in 2002 in Hong Kong. Through categorizing these records, a model with six components (i.e., ideas, design, interpretation, evaluation, understanding and application, and evaluation) was proposed to illustrate the cognitive understanding and the practical processes that are necessary for conducting scientific inquiry in primary school. These six components were further reorganized into two domains (i.e., process and outcome) in the model proposed by So (2006) in her investigation of written project reports and oral presentations of 16 groups in another primary science project event.

Apart from the efforts to generate a framework that can be used to theorize scientific inquiry in primary school at the initial stage of promoting scientific inquiry in Hong Kong, other studies have probed various issues in the process of conducting scientific inquiry, such as pupils' cognitive understanding and reasoning ability required for designing scientific inquiry (Lee and Ng 2004), collaborative learning environments (So and Ching 2011), mathematical skills (So 2013), and representation practice (So 2014). Lee and Ng (2004) used an inquiry task on heat conduction to explore the extent to which primary pupils' cognitive understanding and reasoning ability was ready for doing scientific inquiry. Lesson observation and the analysis of their worksheets revealed that although they had some understanding of the need to control variables, there were considerable limitations to pupils' understanding which prevented them from designing a reliable and valid investigation. It was also found that primary 5 pupils had a better understanding of and reasoning ability for conducting scientific inquiry than their primary 4 counterparts.

So and Ching (2011) designed a collaborative science learning environment to help teachers incorporate small-group learning with science inquiry. The survey of 647 primary pupils and interviews with 47 teachers indicated that the use of group roles made the division of work among group members more even, and reduced the conflicts among pupils, and so involved pupils more effectively in the scientific inquiry task. Teachers still need to help pupils to become familiar with the complex cooperative structures in a progressive manner.

Within this empirical context, science and mathematics come together to solve problems with evidence, construct knowledge claims, communicate claims, and persuade others that the claims are valid and useful. So (2013) explored the mathematics skills used in 26 extracurricular open science inquiries, with the results revealing that pupils used measurement, numeracy skills of counting and calculation, and tables and graphs in their science inquiries. Although most measurements in the science inquiries were used appropriately, there is room for improvement with other mathematical procedures that involve higher-level thinking skills, such as analyzing and calculating numerical data and interpreting graphs and tables.

During science inquiry, pupils have the opportunity to express their knowledge in a wider range of forms. So (2014) investigated the use of representation of 145 science inquiry project entries in a primary science exhibition in Hong Kong, of which 22 were awarded Outstanding, 46 were awarded Merit, and 77 were awarded Consolation. It was found that the highest-level projects (judged Outstanding) tended to exhibit a higher frequency and more diverse use of representations. In addition, the three award levels also displayed different patterns of representational practices.

Yeung (2009) analyzed the perception and implementation of inquiry in teaching of the teachers in two Hong Kong primary schools. It was found that the teachers mostly expressed positive perceptions about inquiry-based teaching and learning. However, there were gaps between teachers' perceptions and practices. Limited experience and confidence to lead students to inquiry among the teachers would be one important impeding factor. It is recommended that continuous supports by the government to improve the local culture in teaching and the school curriculum should be needed so as to promote inquiry teaching in primary schools.

2.5.3 Science Teachers and Their Development

Primary science teachers and their development are another important area of research in Hong Kong. Hong Kong scholars have explored teachers' cognitive complexity when they design science lessons, the features of their development, as well as lesson study and university-school partnership as the strategies for teacher professional development.

Through adopting the five-level Structure of Observed Learning Outcomes (SOLO) taxonomy (i.e., from pre-structural, through uni-structural, multi-structural, relational, to extended abstract), So (1997) investigated 10 in-service primary science teachers' cognitive complexity when planning science lessons. After analyzing the data collected by process tracing, journal keeping, thinking aloud, interviews, and simulated recall, significant differences were found in the levels of structural complexity in teacher cognition among teachers who had science teacher education and those who did not. For the teachers with science education training, more relational responses were identified, which implied that their thinking tended to be dense and richly interconnected in a highly structured way, and they were more capable of relating different aspects of teaching and learning when planning science lessons.

Besides the status of teachers' cognition, So and Watkins (2005) conducted a longitudinal study to explore how 25 teachers' thinking changed from pre-service teacher education through to their first year as classroom teachers. On the basis of the findings of four major aspects, including conceptions, planning, teaching, and reflection, the authors found that the majority of the teachers became more constructivist in terms of their conceptions and teaching practice, but they also tended to become more simplistic in planning and less coherent in thinking as they progressed from pre-service teacher education to beginning teachers in schools.

The lesson study approach, which was developed on the basis of variation theory (Marton 1986) and action research, has long been adopted in Hong Kong to facilitate pre-service and in-service teachers' professional development. Variation theory focuses on the object of learning and explores "how a learner might come to see, understand, or experience a given phenomenon in a certain way" (Orgill 2012, p. 3391). Teachers can use variation theory as a guiding principle of pedagogical design for it helps explain the differences in students' learning of a particular concept (Lo and Marton 2011). Two studies have investigated the effectiveness in the context of the primary science classroom. Lo-Fu and So (2004) found that after conducting lesson studies of "Water in Air," "The Formation of a Solar Eclipse," and "Flow of Electricity," teachers made improvements in their teaching after experiencing the "planning-teaching-evaluation" process. Significant progress was also identified in pupils' achievement in the pre- and posttests. Similar improvement in pupils' achievement was also observed for the topic of the color of light (Lo et al. 2006). It was argued that teachers can make use of the variation theory framework to analyze their own teaching practice to develop an analytical awareness of teaching and learning.

2.5.4 Information Technology in Science Education

Rapid advancements in IT and the government's efforts to promote the use of IT in education have stimulated the practice and research interest in applying information technology both in school science learning and science teacher training.

IT in the Primary Science Classroom

A line of works has explored the use of multimedia resources in primary classrooms. So and Kong (2010) designed a multimedia learning unit that helps learners understand the natural phenomenon of the Earth's movement. The unit was taught by a teacher in two classes with different pedagogical approaches (i.e., the learner-centered and teacher-centered approaches). The pre- and post-lesson tests showed significant improvements in learning achievement in both classes, but the learner-centered approach resulted in better achievement. Kong and So (2008) adopted the concept of a resource-based learning environment (RBLE) to develop pupils' understanding of family trees. RBLE is comprised of three components, an audiovisual database for guided and coupled inquiry, an interactive interface for conceptualizing the nomenclature, and a tool for learners to construct their own family trees. The interview findings and a comparison of the pre- and posttest indicated that RBLE can assist learners in building knowledge of family trees. A positive outcome was also reported in a study of applying RBLE to the unit of learning animal classification (So and Ching 2015), regardless of the teaching approach being deductive or inductive.

Another study (So and Kong 2010) investigated the interaction of pupils' academic background and support levels in a RBLE on Earth movement. It found that RBLE provided little support to pupils of lower academic background, but was effective for those of higher academic background. The views of teachers on the use of online resources to support inquiry learning were probed in So et al. (2012). It was found that traditional Web 1.0 resources are more commonly used, and when these online resources are used, teachers take the more dominant role in using them to support lecturing and explaining. In addition, some associations between inquiry practices and use of online resources were observed. After investigating how five teachers planned to integrate online resources in their teaching of science topics in the primary curriculum, So (2012) generated a resource-based e-learning environment (RBeLEs) framework that can be used as a planning tool for teachers who wish to integrate the Internet into their classroom teaching.

Besides multimedia resources used in the primary science classroom, other forms of applying IT have also been explored in Hong Kong. Cheng and Li (2002) studied three IT-oriented approaches to science learning (i.e., data logger experiments, web technologies in a science project, and use of IT in science assessment) and found that they all enhanced the pupils' science learning. Yuen (2005) engaged primary pupils in an interschool asynchronous online threaded discourse through a computer-mediated communications platform. The results showed that online discourse could broaden the basis for learning and teaching science. Li et al. (2006) discovered that the use of the cognitive perturbation strategy in conjunction with a dynamic computer-supported modelling environment could facilitate the conceptual change of pupils. Kong et al. (2009) found the remote-controlled experiments helped senior primary school students learn electrical circuits more effectively. Song and Kong (2014) and Song (2014) investigated how the science learning was supported by mobile devices. It was found that when mobile learning was incorporated into scientific inquiry, pupils advanced their domain knowledge and developed their inquiry skills.

Video-Based Learning Environment

The published research of applying information technology in science teaching training focuses on how to make use of a digital video database to facilitate the development of pre-service science teachers. Cheng et al. (2005) reported a website to make use of video clips on exemplary parts of pupils' teaching performance in schools or microteaching to support their student teachers' development of teaching competence during various kinds of field experience activities such as practicums, teaching practice, school visits, etc. The website also provided a platform for pre-service teachers to share their lesson plans, schemes of work, and other teaching resources. As indicated in the comments and discussion posted on the website, these video clips provided and promoted alternative teaching strategies. They served to demonstrate certain teaching skills and approaches, stimulate pupils' reflections on critical incidents, start debates on controversial issues related to science teaching, and allow criticism of problematic teaching.

Three papers (So et al. 2008, 2009; So 2012) reported another platform to utilize an online digital video database for developing pre-service teachers, to which the teachers needed to record, edit, and upload their own teaching practice videos for online peer evaluation. A Video Communication Panel and Bulletin Board were incorporated into the platform to facilitate discussion of the clips by members of the learning community. This platform was first piloted with small groups of teachers (three in So et al. 2008; four in So et al. 2009). Content analysis of the sharing in the discussion forum and individual teacher reflections indicated that the involvement in this learning community helped them develop better understanding of teaching techniques and good teaching practices. Both the video provider and the viewers benefited from the flow of suggestions and feedback among the members of the learning community. The difficulties mentioned were mainly technical problems encountered in viewing video segments and the shooting of lessons. So (2012) applied this platform with a larger number of pre-service teachers ($N = 25$) and investigated the teachers' learning outcomes in the online video-based learning community. Drawing from the Structure of Observed Learning Outcomes (SOLO) taxonomy, the author analyzed the teachers' written comments and feedback recorded in the platform and the post-course interviews and found that the majority of comments and feedback were classified as uni-structural. The interviews revealed that the student teachers benefited from the opportunities to engage in peer interaction and self-reflection.

2.5.5 Environmental Education

Environmental education may be the youngest branch of research in primary science education in Hong Kong. The few recent studies conducted including So et al. (2014) examined the impacts of the inquiry learning strategies employed in a "Plastic Education Project" on primary pupils' knowledge, beliefs, and intended behavior. It was found that the inquiry-oriented strategies significantly improved the pupils' knowledge of the types of plastic waste and their beliefs about plastic waste. However, these strategies might not be effective in changing pupils' intended behavior regarding plastic waste classification and management. Cheng and So (2015) surveyed primary teachers' environmental literacy. It was found that although individuals were able to score highly on environmental literacy, others' scores were weak, and some even did not show much concern at all for the environment. The difference was discussed in terms of their epistemological beliefs and commitment to environmental education. Chow et al. (2016) introduced the design of a new eight-compartment plastic waste collection bin based on scientific evidence for facilitating plastic recycling and sorting at source. It was found that the volumes of the different kinds of plastic waste collected matched the design dimensions.

2.6 Future Directions for School Science Education

Changes and developments in science education have been witnessed in Hong Kong as in the rest of the world. When Hong Kong was a British colony, the science curriculum was mostly based on the British science curriculum (Wu 1993). Since the 1990s, as the British influence in Hong Kong decreased and a new millennium approached, the science education curriculum began to reflect more the needs of Hong Kong pupils to fully unleash their potential in innovation stated in the 2015 Policy Address by the Hong Kong Government[18] while drawing on the experiences and science education development initiatives of other countries. In response to this, future research on STEM education, IT in science learning, and interdisciplinary approach in science and environmental education that would have the great potential to enhance the quality of primary science education are highly recommended.

2.6.1 STEM Education

As is reported in "The Ecosystem of Innovation and Technology in Hong Kong"[19] by Our Hong Kong Foundation which is a nongovernment, nonprofit organization dedicated to promoting the long-term and overall interests of Hong Kong through public policy research, advocacy, and engagement, "riding on its competitive edge as an international financial hub, Hong Kong needs to develop an innovation and technology strategy to ensure its sustainable development. According to the World Economic Forum 2015–16 Global Competitiveness Report, Hong Kong ranks a respectable No.7 in overall competitiveness, but fares significantly lower in the metrics of innovative capacity, and availability of scientists and engineers." In terms of human resources, in Hong Kong, career prospects of science, technology, engineering, and mathematics (STEM) graduates are considered less attractive than professions such as business and finance, let alone medicine and law. This creates a vicious cycle of weak demand for and supply of scientists and engineers.

In the 2015 Policy Address by the Hong Kong Government[20], it was announced that the Education Bureau will "renew and enrich the curricula and learning activities of Science, Technology and Mathematics (STEM), and enhance the training of

[18] 2015 Policy Address http://www.policyaddress.gov.hk/2015/eng/pdf/PA2015.pdf

[19] ST_report_eng_online.pdf

[20] 2015 Policy Address http://www.policyaddress.gov.hk/2015/eng/p150.html

teachers, thereby allowing primary and secondary pupils to fully unleash their potential in innovation" (Item 152 in 2015). There was more commitment from the government in 2016 Policy Address[21] that "The Government will step up efforts to promote STEM (Science, Technology, Engineering and Mathematics) education and encourage students to pursue the study of these subjects" (Item 89 in 2016). A commitment in 2017[22] Policy Address with "the Education Bureau should strive to promote STEM education with the provision of additional resources for primary schools at the beginning of last year and be prepared to provide each public sector secondary school with an additional one-off subsidy of $200,000 to promote the implementation of school-based programs related to STEM education" (Para. 212 in 2017).

In response to the policy addresses and the report on promotion of STEM education (Education Bureau 2016), there are various initiatives from different stakeholders, including the Education Bureau, education associations, NGOs, teacher educators, and the business sector proposing different strategies to help with STEM education. However, with the different agenda behind the initiatives, their effectiveness is in question. The exploration of strategies to strengthen pupils' integrative learning and application skills can be a possibility by researching pupils' learning experience through the successful science events and activities which have a long history with some baseline data and information for more realistic development. The recent publication based on an analysis of primary students' STEM projects with a researcher-generated framework delineated the pattern of STEM activities in students' work (So et al. 2017) provided a solid foundation for research studies with STEM education.

2.6.2 IT in Science Learning

With the advancements of information and communication technology (ICT), e-learning has become a focus in education with ever-growing popularity. Due to its flexibility and its ability to extend the classroom virtually with the aid of technology, e-learning has become a trend in the education of different countries. The TIMSS report in 2017 stated that 13 out of the 36 participating countries and 3 out of the 7 benchmarking participants had policies on computer use as part of their science curriculum in the fourth grade (Olson et al. 2008). The report discussing the results of national measures taken by 16 of the European Schoolnet's member countries also stated that all the countries generally agree that ICT could diversify the learning process and make the study of mathematics and science and technology subjects more attractive to pupils (Kearney 2010). It was also found in the survey that ICT is used in the teaching of all mathematics and science and technology subjects to varying extents in all of the countries surveyed (Kearney 2010).

[21] 2016 Policy Address https://www.policyaddress.gov.hk/2016/eng/p87.html

[22] 2017 Policy Address https://www.policyaddress.gov.hk/2017/eng/p210.html

Yet, how IT can be effectively incorporated in teaching and learning of primary science is another area of potential development. So (2016), in her presentation of opportunities for e-learning in science education, introduced the major approaches including the evidence-based approach for the topic of "the seasons," the experiential approach for the topic of "the movement of the earth," and the predict-observe-explain approach for the topic of "invisible light." However, the advocacy of IT in education, including the idea of a flipped classroom which inverts traditional teaching methods, delivering instruction online outside of class, and moving "homework" into the classroom, may be very different from teachers' usual practice, and the use of different equipment to achieve the various learning objectives in the science curriculum is costly. Since there is no single "best practice" or "right way" to integrate technology into the classroom, it is worth further investigation to provide the most effective learning environments for pupils.

2.6.3 Interdisciplinary Approach in Science and Education for Environmental Sustainability

There is a growing need for citizens to be involved in the decision-making processes concerning controversial scientific-environmental issues required for sustaining a vibrant community. Are big ideas – climate change, natural disasters, animal adaptation, and energy saving – out of the reach of primary pupils? There are many benefits to learning in a school setting such as improving pupils' interpersonal skills and developing their knowledge, environmental sensitivity, and a sense of belonging to nature (Mobley et al. 2009) which are responsible for shaping pro-environmental behaviors. Although an interdisciplinary approach to science and environmental education is an option to help pupils better understand the knowledge and enhance their attitudes and skills, teachers' experience and training are vital for the execution and implementation of environmental education in schools (Green et al. 2015). Recent studies have also shown that there are often misinterpretations and disconnections in the knowledge of teachers (Colwell and Enderson 2016; Hull et al. 2016), so much effort is needed to support teachers with the necessary knowledge, skills, and attitudes, as well as appropriate teaching strategies to better understand the science in education for sustainability.

2.6.4 Conclusion

As it is emphasized in the report "The Ecosystem of Innovation and technology in Hong Kong"[23] by the "Our Hong Kong Foundation"[24], innovation and technology have been two of the major drivers of global economic growth since the dawn of the

[23] http://ourhkfoundation.org.hk/images/innovation_report/ST_report_eng_online.pdf

[24] http://ourhkfoundation.org.hk/index.php/2015-04-25-06-32-03/2015-04-25-06-32-50

Industrial Age. Invention and commercialization of new sources of power (from steam to electricity) and new materials (such as steel and plastic), combined with innovative ways of organizing production and businesses, have propelled strong productivity gains in many advanced economies. Hong Kong, with a heavy reliance on financial and real-estate sectors, is lagging behind its neighbors in terms of innovation and technology. Therefore, as the world moves into the information technology and Internet-centric era, against a backdrop of increasing globalization, innovation and technology are becoming ever more important factors for Hong Kong's sustainable development.

It is absolutely not an easy task to promote science learning in a packed primary curriculum. With the HKSAR Government support and public call for a long-term strategy to take a holistic approach in driving innovation and technology, incorporating STEM education, IT in education, and science and environmental studies in the primary schooling would be one of the solutions to the better future of Hong Kong.

References

Advisory Committee on Teacher Education and Qualification (ACTEQ). (2009). *Towards a learning profession: Third report on teachers' continuing professional development.* Hong Kong: Government printer.

CDC (Curriculum Development Committee). (1980a). *Syllabus for primary schools: Health Education.* Hong Kong: Government Printer.

CDC (Curriculum Development Committee). (1980b). *Syllabus for primary schools: Social Studies.* Hong Kong: Government Printer.

Chan, S. L., & Lui, C. W. (1998). Future development and characteristics of science curricula in Hong Kong secondary and primary schools. *New Horizons, 39*, 27–36.

Cheng, M. H. M., & Li, K. M. (2002). The use of information technology in science teaching: Implications on student learning. *Asia-Pacific Forum on Science 1Learning and Teaching, 3*(1), Article 2.

Cheng, N. Y., & Lo-Fu, Y. W. (2002). Interdisciplinary curriculum: A case study of General Studies teaching. In Y. T. Cheng, K. T. Tsui, K. W. Chow, & M. M. C. Mok (Eds.), *Subject teaching and teacher education in the new century: Research and innovation* (pp. 317–348). Hong Kong: The Hong Kong Institute of Education.

Cheng, M. H. M., & So, W. M. W. (2009). Implementing science education in an integrated curriculum at primary level: Threats and opportunities. *Curriculum and Teaching, 24*(1), 57–74.

Cheng, M. H. M., & So, W. M. W. (2011). The infusion of strategies for generating high level thinking into the junior secondary science curriculum. In M. H. M. Cheng & W. M. W. So (Eds.), *Science education in international contexts* (pp. 45–62). Rotterdam: Sense Publishers.

Cheng, N. Y., & So, W. M. W. (2015). Teachers' environmental literacy and teaching – Stories of three Hong Kong primary school teachers. *International Research in Geographical and Environmental Education, 24*(1), 58–79 Retrieved from https://doi.org/10.1080/10382046.2014.967111.

Cheng, M. H. M., So, W. M. W., & Yeung, Y. Y. (2005). An effective web-based approach to support the initial training of science teachers. *Journal of the International Society for teacher education, 9*(1), 13–22.

Chow, S., Cheung, Y., & So, W. M. W. (2016). Research and development of a new waste collection bin to facilitate education in plastic recycling. *Applied Environmental Education & Communication, 15*(1), 45–57.

Colwell, J., & Enderson, M. C. (2016). "When I hear literacy": Using pre-service teachers' perceptions of mathematical literacy to inform program changes in teacher education. *Teaching and Teacher Education, 53*, 63–74.

Curriculum Development Committee. (1981). *Syllabus for primary schools: Primary science*. Hong Kong: Government printer.

Curriculum Development Committee. (1994). *CDC syllabus for general studies (Primary I to VI)*. Hong Kong: Government Printer.

Curriculum Development Committee. (2002a). *Basic education curriculum guide- building on strengths (Primary 1- Secondary 3)*. Hong Kong: Government Printer.

Curriculum Development Committee. (2002b). *Science education: Key learning areas curriculum guide (primary 1-secondary 3)*. Hong Kong: Government printer.

Curriculum Development Committee. (2002c). *General studies for primary schools curriculum guide (Primary 1 - Primary 6)*. Hong Kong: Government Printer.

Curriculum Development Committee. (2011). *General studies for primary schools development guide*. Hong Kong: Government Printer.

Curriculum Development Committee. (2014). *Basic education curriculum guide – to sustain, deepen and focus on learning to learn (Primary 1–6)*. Hong Kong: Government Printer.

Curriculum Development Committee. (2017). *General studies curriculum guide for primary schools (P1–P6)*. Hong Kong: Government Printer.

EC (Education Commission). (1990). *Education commission report no. 4: The curriculum and behavioural problems in schools*. Hong Kong: Government Printer.

Education Bureau. (2016). *Report on promotion of STEM education – Unleashing potential in innovation*. http://www.edb.gov.hk/attachment/en/curriculum-development/renewal/STEM%20Education%20Report_Eng.pdf

Education Department. (1967). *Suggested syllabuses for primary schools: Nature study*. Acting government printer at the government press.

Green, C., Medina-Jerez, W., & Bryant, C. (2015). Cultivating environmental citizenship in teacher education. *Teaching Education, 27*(2), 1–19.

Harlen W. (2008). *Science as a key component of the primary curriculum: A rationale with policy implications*. Perspectives on Education 1 (Primary Science), 4–18. Retrieved from www.wellcome.ac.uk/perspectives

Holbrook, J. (1990). Science Education *in Hong Kong: Achievements and determinants* (Education Paper 6). Hong Kong: University of Hong Kong.

Hui, S. C. M. (2006). Benefits and potential applications of green roof systems in Hong Kong, In *Proceedings of the 2nd megacities international conference 2006*, 1–2 December 2006, Guangzhou, China, pp. 351–360.

Hui, S. C. M. (2011). *Technical guidelines for green roofs systems in Hong Kong*. Hong Kong: PRC.

Hull, D. M., Booker, D. D., & Näslund-Hadley, E. (2016). Teachers' self-efficacy in Belize and experimentation with teacher-led math inquiry. *Teaching and Teacher Education, 56*, 14–24.

Isaacs, N. (1962). The case for bringing science into the primary school. In W. H. Perkins (Ed.), *The place of science in primary education*. London: British Association for the Advancement of Science.

Kearney, A. T. (2010). *The urban elite: The A.T. Kearney Global Cities Index 2010*. Retrieved from https://www.atkearney.com/documents/10192/178350/urban_elite-gci_2010.pdf/30bebf01-9591-46ef-8d06-a44b4e5ab364

Kong, S. C., & So, W. M. W. (2008). A study of building a resource-based learning environment with the inquiry learning approach: Knowledge of family trees. *Computers & Education, 50*(1), 37–60.

Kong, S. C., Yeung, Y. Y., & Wu, X. Q. (2009). An experience of teaching for learning by observation: Remote-controlled experiments on electrical circuits. *Computers and Education, 52*(3), 702–717.

Lee, C. K. J. (1998). Primary school teachers' receptivity to curriculum changes in Hong Kong: A comparison between target-oriented curriculum and general studies. *Curriculum Forum, 7*(2), 71–83.

Lee, Y. C., & Kwok, P. W. (2009). Can an egg-dropping race enhance students' conceptual understanding of air resistance? *Physics Education, 44*(2), 151–158.

Lee, Y. C., & Ng, P. H. (2004). Hong Kong primary pupils' cognitive understanding and reasoning in conducting science investigation: A pilot study on the topic of "keeping warm". *Asia-Pacific Forum on Science Learning and Teaching, 5*(3), Article 8.

Li, S. C., Law, N., & Lui, K. F. A. (2006). Cognitive perturbation through dynamic modelling: A pedagogical approach to conceptual change in science. *Journal of Computer Assisted Learning, 22*(6), 405–422.

Lin, F. (2015). *Fostering scientific epistemology among elementary-school students through knowledge building* (Doctoral dissertation). Retrieved from The HKU Scholar Hub.

Lo, S. Y. (2002). *Explaining the development of the primary science curriculum in Hong Kong*. Unpublished thesis.

Lo, T. Y., & Cheng, N. Y. (2001). A study of the Hong Kong primary general studies as an integrated curriculum: Theory and practice. *New Horizons in Education, 44*, 52–63.

Lo, M. L., & Marton, F. (2011). Towards a science of the art of teaching: Using variation theory as a guiding principle of pedagogical design. *International Journal for Lesson and Learning Studies, 1*(1), 7–22.

Lo, M. L., Chik, P. M. P., & Pang, M. F. (2006). Patterns of variation in teaching the colour of light to primary 3 students. *Instructional Science, 34*(1), 1–19.

Lo-Fu, Y. W., & So, W. M. W. (2004). Lesson studies at primary science classrooms: "Water in the air", "solar eclipse", "flow of electricity". *Asia Pacific Forum on Science Learning and Teaching, 5*(3), Article 3.

Marton, F. (1986). Phenomenography: A research approach to understanding different understandings of reality. *Journal of Thought, 21*, 28–49.

Mobley, C., Vagias, W. M., & DeWard, S. L. (2009). Exploring additional determinants of environmentally responsible behavior: The influence of environmental literature and environmental attitudes. *Environment and Behavior, 42*(4), 1–28.

O'Donnell, S. (2002). *International review of curriculum and assessment frameworks – Comparative tables and factual summaries – 2002*. England: Qualifications and Curriculum Authority and National Foundation for Educational Research.

O'Donnell, S., & Micklethwaite, C. (2000). *International review of curriculum and assessment frameworks – Thematic probe: Science for the 21st century, Queensland (Australia), Ontario (Canada), France, the Netherlands and Sweden*. England: Qualifications and Curriculum Authority and National Foundation for Educational Research.

Olson, J. F., Martin, M. O., & Mullis, I. V. (Eds.). (2008). *TIMSS 2007 technical report*. IEA TIMSS & PIRLS.

Orgill, M. (2012). Variation theory. In N. M. Seel (Ed.), *Encyclopedia of the sciences of learning* (pp. 3391–3393). Heidelberg: Springer-Verlag GmbH.

So, W. M. W. (1997). A study of teacher cognition in planning elementary science lessons. *Research in Science Education, 27*(1), 71–86.

So, W. M. W. (Ed.). (1998). *Inquiry activities for Hong Kong primary schools: theory and practice*. Hong Kong: Hong Kong Institute of Education.

So, W. M. W. (2002). Science learning at the primary level: Constructivist learning and integration of learning experiences in curriculum development. *Journal of Basic Education, 11*(2), 35–53.

So, W. M. W. (2003). Learning science through investigations: An experience with Hong Kong primary school children. *International Journal of Science and Mathematics Education, 1*(1), 175–200.

So, W. M. W. (2006). Inquiry into primary pupils' science projects: Implication for design of inquiry learning. *Asia-Pacific Forum on Science Learning and Teaching, 7*(1), Article 3.

So, W. M. W. (2008). Primary science learning of Hong Kong through the lens of TIMSS (從TIMSS透視香港的小學科學學習). *Asia-Pacific Forum on Science Learning and Teaching, 9*(1), Article 16.

So, W. M. W. (2012). Quality of learning outcomes in an online video-based learning community: Potential and challenges for student teachers. *Asia-Pacific Journal of Teacher Education, 40*(2), 143–158.

So, W. M. W. (2013). Connecting mathematics and science in primary science inquiry. *International Journal of Science and Mathematics Education, 11*(2), 385–406.

So, W. M. W. (2014). Representational practices in extra-curricular science inquiry projects: A study with Asian primary pupils. *International Journal of Science and Mathematics Education, 14*(1), 55–79.

So, W. M. W. (Chief Ed.). (2015). *Go green at home*. Hong Kong: The 18th Organizing Committee of Innovations in Science and Environmental Studies.

So, W. M. W. (2016). Opportunities for e-learning in science education. Presentation at the 2016 KASE International Conference "Future Science Education: International perspectives" *Kyungpook National University* Daegu Metropolitan City, Korea 28–30 January, 2016.

So, W. M. W., & Cheng, M. H. M. (2000). A study on the professional development of primary school teachers in teaching science. *Journal of Science and Mathematics Education in Southeast Asia, 23*(2), 1–18.

So, W. M. W., & Ching, N. Y. F. (2011). Pupil science learning in resource-based e-learning environments. *Journal of Computers in Mathematics and Science Teaching, 30*(2), 203–223.

So, W. W. M., & Ching, F. N. Y. (2015). A study of the internet resource-based e-learning environments in Hong Kong: Animal classification in a primary school. In *Science education in East Asia* (pp. 317–337). Cham: Springer International Publishing.

So, W. M. W., & Kong, S. C. (2010). Interaction of student's academic background and support levels in a resource-based learning environment on Earth's movement. *Interactive Learning Environments, 18*(2), 153–176.

So, W. M. W., & Watkins, D. (2005). From beginning teacher education to professional teaching: A study of the thinking of Hong Kong primary science teachers. *Teaching and Teacher Education, 21*(5), 525–542.

So, W. M. W., Cheng, M. H. M., & Tsang, C. L. (1998a). Problems of teaching science-related topics in Hong Kong primary school. *Journal of Basic Education, 7*(2), 43–57.

So, W. M. W., Cheng, M. H. M., & Tsang, C. L. (1998b). Difficulties encountered by primary general studies teachers in teaching science-related topics. *Journal of Basic Education, 7*(2), 43–57.

So, W. M. W., Cheng, M. H. M., Leung, C. Y., & Wong-Yu, L. W. (1999). The teaching of general studies as an "integrated" subject in Hong Kong primary schools. *Journal of Basic Education, 8*(2), 13–28.

So, W. M. W., Hung, H. K. V., & Yip, Y. W. W. (2008). The digital video database: A virtual learning community for teacher education. *Australasian Journal of Educational Technology, 24*(1), 73–90.

So, W. M. W., Pow, J., & Hung, V. (2009). The interactive use of a video database in teacher education: Creating a knowledge base for teaching through a learning community. *Computers & Education, 53*(3), 775–786.

So, W. M. W., Cheng, M. H. M., Kong, S. C., & Ching, N. Y. F. (2012). Views of primary science teachers towards the use of online resources to support the implementation of inquiry learning. *The Education, 3–13*, 1–16.

So, W. M. W., Cheng, N. Y. I., Chow, C. F. S., & Zhan, Y. (2014). Learning about the types of plastic wastes: Effectiveness of inquiry learning strategies. *Education, 3–13*, 311–324.

So, W. M. W., Zhang, J., Chow, C. F., & Leung, C. F. (2017). Analysis of STEM activities in primary science projects in an informal learning environment. *International Journal of Science and Mathematics Education*. https://doi.org/10.1007/s10763-017-9828-0.

Song, Y. (2014). "Bring your own device (BYOD)" for seamless science inquiry in a primary school. *Computers & Education, 74*, 50–60.

Song, Y., & Kong, S. C. (2014). Going beyond textbooks: A study on seamless science inquiry in an upper primary class. *Educational Media International*, Online publication, 1–10.

Tucker, M. (2016). Hong Kong: In the crosshairs of the future. *Eductation Week*. http://blogs.edweek.org/edweek/top_performers/2016/06/hong_kong_in_the_crosshairs_of_the_future.html.

Wu, K. B. (1993). *Science and technology education in Taiwan (China)*. Washington, DC: World Bank.

Yeung, S. Y. S. (2009). Is student–centered pedagogy impossible in Hong Kong? The case of inquiry in classrooms. *Asia Pacific Education Review, 10*, 377–386.

Yuen, H. K. (2005). Learning science through online threaded discourse. In D. Hung & M. S. Khine (Eds.), *Engaged learning with emerging technologies* (pp. 127–147). Dordrecht: Springer.

Chapter 3
Elementary Science Education in Japan

Kenji Matsubara

General Information

The current Constitution of Japan was established after World War II and stipulated our basic national educational policy as follows: "All people shall have the right to receive an equal education corresponding to their ability, as provided by law. The people shall be obligated to have all boys and girls under their protection receive ordinary education as provided for by law. Such compulsory education shall be free" (Article 26). In accordance with the spirit of the constitution, the Basic Act on Education was drafted and put into effect in March 1947 (and amended in 2006). This stipulated the aims and principles of education: Article 1 stated that education shall aim for the full development of personality and strive to nurture the citizens, sound in mind and body, who are imbued with the qualities necessary for those who form a peaceful and democratic state and society. These ideals therefore form the major underlying principles of post-World War II education in Japan that provides students with equal opportunities to receive 9 years of compulsory but free education.

At present, Ministry of Education, Culture, Sports, Science, and Technology (MEXT) is responsible for school education in Japan, and all educational activities come under its supervision. MEXT supervises and subsidizes prefectural or municipal boards of education. Then, these boards of education establish and take care of the running of virtually all elementary schools. Both public and private institutions therefore serve primary education; private schools are basically self-supporting entities funded through tuition fees, donations, and contributions from business

K. Matsubara (✉)
National Institute for Educational Policy Research (NIER), Ministry of Education, Culture, Sports, Science and Technology (MEXT), Tokyo, Japan
e-mail: kmatsuba@nier.go.jp

Y.-J. Lee, J. Tan (eds.), *Primary Science Education in East Asia*, Contemporary Trends and Issues in Science Education 47,
https://doi.org/10.1007/978-3-319-97167-4_3

organizations, while national and prefectural (local) governments still provide financial assistance toward maintaining and improving facilities of private schools. According to the Basic Survey on Schools, 1.1% of elementary school students are enrolled in private schools throughout Japan (MEXT 2014a).

The education system in Japan follows a six–three–three pattern: 6 years of primary education (International Standard Classification of Education [ISCED] 1), 3 years of lower secondary education (ISCED2), and 3 years of upper secondary education (ISCED3). For primary and lower secondary education, which are compulsory, almost all children aged 6–15 are enrolled in school. The data from 2014 shows that 98.4% of this age group went on to upper secondary school and 51.5% entered a university (MEXT 2014b). In public elementary schools, there is no official policy on within-school streaming nor are public primary students tracked. From the 3rd grade of primary education, a compulsory program of science is taught to all students in mixed-ability classes where the same curriculum is prescribed for all students. For the classes or schools for students with special needs, a specially adapted curriculum is available.

As is widely known, the majority of the inhabitants in this country are Japanese, and the language of instruction in school is Japanese. Recently, however, the number of foreign national residents (medium- or long-term residents and special permanent residents) has gradually increased, accounting for 1.6% of all residents in Japan (Statistics Bureau and Statistical Research and Training Institute 2016). The population includes Chinese (0.5%), Korean (0.4%), Brazilians (0.1%), and other minority groups. In some regions, having Brazilian communities, education is provided both in Japanese and Portuguese, for instance.

Course of Study

The Japanese national curricula are known as "Courses of Study." The Course of Study for Elementary Schools (COSES) accommodates the national curriculum for primary education, while the Course of Study for Junior High Schools caters for lower secondary education. The primary science curriculum is included in the COSES, and it has been revised eight times since its implementation in 1947. The Course of Study for Science started as a "tentative plan" in 1947 and was regarded as guidance for teachers conducting research. The goal of science education and teaching described in the 1947 Course of Study was "for children and pupils to foster the following three qualities related to problems in their environment so that everyone can have a rational life style and a better living." The three qualities refer to:

1. Abilities to observe, think about, and treat things scientifically
2. Knowledge about the principles and applications of science
3. Attitudes of seeking truth and being willing to create new things

The content, which was for all grade levels, comprised of topics about (1) animals and humans, (2) plants, (3) the inorganic environment, (4) machines and tools, and (5) health.

After the war, the Ministry of Education advocated the fostering of scientific knowledge, ability, and attitude and indicated the importance of creative learning for children according to their interests and needs and of generalizing knowledge as they collaborate with other children. It emphasized the need to change from teacher-centered to child-centered approaches. The point the Minister of Education made was a paradigm shift from prewar science education to train citizens to one that fosters scientific knowledge, ability, and the attitude to scientifically and logically resolve problems arising in everyday life, contributing to children's later life in society (Shiba 2011).

The 1952 Course of Study in science, as the first revision, had the same basic policy as the 1947 version meaning that pupils were to attempt to resolve problems through a variety of activities and to understand scientific principles and laws. During this period of rapid economic growth, improvement in basic academic performance was strongly promoted. As a result, in the 1958 Course of Study, the child-centered approach was reconsidered and more importance was placed on educational content. The content became more systematized and scientific. With this revision, the focus of the problem-solving learning and life unit science started to shift to that of systematic (methodical) learning. In the 1960s, the modernization movement of education, which was triggered by the Sputnik crisis in the USA, influenced Japan's primary science education as well. By the time the 1968 Course of Study appeared, discovery and inquiry learning had become widely prescribed for teaching science, and by the 1970s, environmental issues came to the forefront, while national economic growth became slower. The 1977 Course of Study spoke of realizing a comfortable and fulfilling school environment, by reducing the burden of study for children. "Comfortable (pressure-free) and fulfillment" was one of the slogans during that time. Some aspects of the media started calling for the emphasis on *Yutori* education or more relaxed education (see https://en.wikipedia.org/wiki/Yutori_education). In science education, child-centered activities with emphasis of direct experiences received more attention.

In the 1989 Course of Study, science as well as social studies was discontinued for Grades 1 and 2, and they were replaced with a new subject called living environment studies. With the implementation of the 1998 COSES and the 5-day-week school system, the science content for Grade 3 to 6 students was reduced by approximately 30%. This newly revised Course of Study aimed to help pupils develop a "zest for living," an ability to independently find, learn, and think about a study topic and to help develop a solid problem-solving ability by further exploring reduced content not only through education at school but through education at home and in the community.

Table 3.1 shows the changes in the number of science classes per week designated by the COSES. As previously mentioned, since the 1989 revision, science for Grades 1 and 2 has been replaced with living environment studies. In the latest revision, enriching and strengthening science was one of the emphases of the curriculum revision, and this led to additional classes for primary science. As shown in the

Table 3.1 Number of science classes per week designated by the Course of Study in Elementary Schools (COSES)

The course of study		Number of science classes per week (the numbers noted in brackets show the number of science classes a year)					
Year published	Year implemented	Grade 1	Grade 2	Grade 3	Grade 4	Grade 5	Grade 6
1947	1947	2	2	2	3	3–4	3–4
1958	1961	2	2	3	3	4	4
1968	1971	2	2	3	3	4	4
1977	1980	2	2	3	3	3	3
1989	1992	–	–	3	3	3	3
1998	2002	–	–	2 (70)	2.6 (90)	2.7 (95)	2.7 (95)
2008	2009	–	–	2.6 (90)	3 (105)	3 (105)	3 (105)

Source: Shiba 2011

table, 10 h was added for Grades 5 and 6 science, 15 h was added for Grade 4 science, and 20 h was added for Grade 3 science on a yearly basis.

3.1 Curriculum Visions, Goals, Emphases, and Structure of Primary Science

The current COSES was announced in March 2008 and fully implemented from April 2011. Some parts of the new curricula for mathematics and science were partially implemented during a transition period from April 2009 to March 2011. The COSES has subject-specific sections that consist of three parts: overall objectives, objectives and content for each grade, and syllabus design and handling the content. The science section of the Course of Study lists the overall objectives of the primary science, the objectives and content for each grade (Grades 3–6), and the syllabus design and handling the content for the primary science. The latter section describes several points to consider when preparing a syllabus. Then, under the section of objectives and contents for each grade for primary science, there are subordinate sections that include objectives, content, and handling [teaching] of the content. The last section elaborates the teaching and learning of content in the unit.

Examples of the overall objectives in primary science together with some of the objectives (grade specific) and content in Grade 4 are shown below.

Overall Objectives

To enable pupils to become familiar with nature and to carry out observations and experiments with their own prospectus, as well as to develop their problem-solving abilities and nurture hearts and minds that are filled with an affection for the natural world, and, at the same time, to develop a realistic understanding of natural phenomena and to foster scientific perspectives and ideas

Grade 4

1. Objectives

   (1) To develop perspectives and ideas about the properties and functions of objects, by investigating air, water, changes in the state of an object, and electrical phenomena, in relation to the functions of power, heat, and electricity, and through probing the identified problem and making learning materials with interest
   (2) To foster an attitude to love and protect living things and to develop perspectives and ideas about the structure of the human body, activities of animals/growth of plants, meteorology, and movement of the moon and stars, by investigating them in relation to movement, seasons, temperature, and time, through probing the identified problems with interest

2. Content

   A. Matter/Energy

      1. Properties of air and water

         To develop pupils' ideas about the properties of air and water by exploring the changes in their volume and pressure in compressing air and water in a closed space

         (a) When air is compressed in a closed space, the volume decreases and the pressure increases.
         (b) Air in a closed space can be compressed, but water cannot be compressed.

   B. Life/the Earth

      1. Structure and movement of the human body

         To develop pupils' ideas about the relationship between the structure and movement of the human body, by exploring the movement of bones and muscles and by observing the movement of humans and other animals or by using teaching materials

         (a) The human body has bones and muscles.
         (b) The human body can move due to the functions of bones and muscles (source: MEXT 2008a).

The learning objectives from each grade belong to either of the two groups: Matter/Energy or Life/the Earth. The grouping was done by taking into account the characteristics of the learning objects and the perspectives and ideas developed by the pupils. Objectives relating to Matter/Energy in each grade include making things/products since the focus here is on authentic understanding. On the other hand, objectives relating to Life/the Earth include developing an attitude of care and protection of living things and to respect life, since this part focuses on nurturing hearts and minds that are filled with an affection for the natural world (MEXT 2008b).

Figures 3.1 and 3.2 show the mapping of the items of scientific knowledge and concepts (content) covered in primary and junior high science. The figures

School Level	Grade	Energy: Perspective on energy	Energy: Conversion and conservation of energy	Energy: Efficient use of energy resource
Elementary school	Grade 3	Function of wind and force of rubber • Function of wind • Force of rubber Properties of light • Reflection and collection of light • Striking of light and its brightness and warmth	Properties of magnet • Objects attracted to a magnet • Opposite poles and like poles Pathway of electricity • Patterns of circuit connections that conduct electricity • Materials that conduct electricity	
	Grade 4		Function of electricity • Number and circuit of dry cell battery • Function of photocell	
	Grade 5	Movement of pendulums • Movement of pendulums ☆	Function of electric currents • Magnetization of an iron core,change of polarity (transit from Grade 6) • Strength of an electromagnet (transit from Grade 6)	
	Grade 6	Regularity of a lever • Balance and weight of a lever (transit from Grade 5) • Regularity of balance of a lever (transit from Grade 5) • Use of a lever (tools using a lever in everyday life)	Use of electricity • Generation and storage of electricity • Transformation of electricity (transformation into light, sound, heat, etc.) • Heat generation by electricity • Use of electricity (tools using electricity in everyday life)	
Junior high school	Grade 7	Power and pressure • Function of power (including power and stretching of springs, difference between weight and mass) • Pressure (including water pressure) Light and sound • Reflection and refraction of light • Function of convex lens • Nature of sound		
	Grade 8	Electric current • Circuit and current/voltage • Current/voltage and resistance • Electric and energy (including the amount of electric energy, heat) • Static electricity and current (including electronic) Current and the magnetic field • Magnetic field created by the currents • Changes in force in the magnetic field • Electromagnetic induction and power generation (including alternating current)		
	Grade 9	Regularity of movement • Balance of forces (transit from Grade 7) (including the synthesis and degradation of the force) • Speed and direction of movement • Force and motion Mechanical energy • Work and energy (including collision (transit from Grade 5), work rate) • Preservation of mechanical energy	Energy • Various energy and its transformation energy transformation efficiency) • Energy resources (including radiation)	Progress • Progress Conservation and utilization • Conservation and utilization Common

Fig. 3.1 Composition of content focusing on "energy" and "particles" in elementary and junior high school science (based on draft English translation)
Solid line indicates a new item. Broken line indicates an item to transit. "☆" mark indicates an item changed from an optional to a compulsory item
Source: MEXT 2008b pp.14–15. The translation is not official

Particles			
Presence of particles	Binding of particles	Preservation of particles	Energy possessed by particles
		Object and weight • Shape and weight • Volume and weight	
Properties of air and water • Air compression • Water compression electricity			Metal, water, air and temperature • Change of temperature and volume • The difference in the way of getting warm • Three changes of state of water
		Dissolution of substances • Limit to the amount of solute that can be dissolved in a solvent • Change of the amount of solute in a solvent • Conservation of weight	
Mechanism of combustion • Mechanism of combustion	Properties of aqueous solutions • Acid, alkaline and neutral aqueous solutions • Aqueous solutions with gas dissolved inside • Aqueous solutions that change metals		
Appearance of substances • Familiar substances and its nature (including plastic) • Generation and the nature of gas		Aqueous solution • Dissolution of substance • Solubility and recrystallization of gas	Change in state • State change and heat • Melting point and boiling point of a substance
Origin of substances • Decomposition of substances • Atoms and molecules re of gas	Chemical changes • Chemical combination • Oxidation and reduction (transit from Grade 9) • Chemical changes and heat (transit from Grade 9) Chemical changes and mass of substance • Chemical changes and conservation of mass • Regularity of mass change		
Aqueous solution and ion • Electrical conductivity of aqueous solution • Origin of an atom and an ion • Chemical change and battery	Acid/alkali and ion • Acids and alkalis (transit from Grade 7) • Neutralization and salt (transit from Grade 7)		
(including heat conduction and in scientific technology in scientific technology ☆ of the natural environment of scientific technology of the natural environment of scientific technology with Field 2>			

Fig. 3.1 (continued)

School Level	Grade	Life			
		Structure and function of living things	Diversity and commonality of living things	Continuity of life	Relationship between living things and environment
Elementary school	Grade 3	Insects and plants • Growth and body structure of insects • Growth and body structure of plants			Observation of familiar environments • Conditions of familiar living things • Familiar living things and their interactions with their surrounding environment
	Grade 4	Structure and function in the human body • Bones and muscles • Functions of bones and muscles (including functions of joints)	Seasons and living things • Activities of animals and seasons • Growth of plants and seasons		
	Grade 5			Germination, growth and fruiting of plants • Nutrition in seeds • Conditions of germination • Conditions of growth • Pollination and fruiting of plants Birth of animals • Growth inside eggs☆ • Small living things in water • Growth inside the mother☆	
	Grade 6	Structure and movement of the human body • Respiration • Digestion and absorption • Blood circulation • Presence of major organs (lung, stomach, small intestine, large intestine, liver, kidney, heart) Nutrition of plants and pathway of water • Production of starch • Pathways of water			Living things and the environment • Relationship of living things with water and air • Relationship of living things through food
Junior high school	Grade 7	Structure and function of plants • Structure and function of flowers • Structure and functions of leaves, stems, roots	Group of plants • Group of seed plants • Group of plants that do not make seeds		Observation of living things • Observation of living things
	Grade 8	Structure and function of animals • Functions to maintain life • Stimulus and response	Living things and cells • Living things and cells (transit from Grade 9) Group of animals • Group of vertebrates • Group of invertebrate History and evolution of living things • History and evolution of living things		
	Grade 9			Growth and proliferation of living things • Cell division and growth of living things • Proliferation of living things Regularity of inheritance and gene • Regularity of inheritance and gene (including DNA)	Living things and the • Balance in the natural • Study of natural environment (including Benefits and disasters • Benefits and disasters Preservation of natural scientific technologies • Preservation of scientific technologies <Common with Field 1>

Fig. 3.2 Composition of content focusing on "life" and "the Earth" in elementary and junior high school science
Solid line indicates a new item. Broken line indicates an item to transit. "☆" mark indicates an item changed from a selectable item to a compulsory item
Source: MEXT 2008b pp.16–17. The translation is not official

The Earth		
Interior of the Earth	Surface of the Earth	Surrounding of the Earth
	The sun and the ground • Position of shady spots and movement of the sun • Differences in warmth and dampness of the ground	
	Weather conditions • Change in temperature in a day depending on weather (transit from Grade 5) • Natural evaporation and condensation of water	The moon and stars • Shape and position of the moon • Levels of brightness and colors of stars • Position of stars
Function of running water • Function of running water (cutting, transportation, piling up materials) • Upstream, downstream, and pebbles on a riverside • Rain and swelling of water	Weather change • Cloud and weather change • Forecast of weather change	
Formation and change of land • Formation of land and expansion of geographical strata • Formation of geological strata and fossil remains • Volcanic eruptions and land change by volcanoes ☆		The moon and the sun • Positions and shapes of the moon and position of the sun • Condition of the moon's surface
Volcano and earthquake • Volcanic activities and igneous rocks • Transmission of earthquake and function of interior of the Earth Overlap of geological strata and past conditions • Overlap of geological strata and past conditions		
	Weather observation • Weather observation Weather changes • Generation of fog and clouds • Passing of the fronts and weather changes Weather in Japan • Weather features in Japan • Movement of air and influence of the ocean	
environment world environment and preservation of global warming, alien species) of nature of nature ☆ environment and utilization of natural environment and utilization of		Movement of astronomical bodies and rotation/revolution of the Earth • Daily motion and rotation • Annual motion and revolution The solar system and the stars • State of the sun • Movement and appearance of the moon (including solar eclipse, lunar eclipse) • Planets and stars (including the presence of a galaxy)

Fig. 3.2 (continued)

indicate what content is taught in which grade as well as progressions of learning. The groups of learning content related to each of disciplines, such as physics, are described as pillars, such as energy. The pillars can function to emphasize follow-up and connections of scientific knowledge and concepts in each of the disciplines through Grades 3 to 9. One might want to consider the pillars as big ideas as they are key ideas in each discipline. The figures also intend to help teachers easily grasp what content is newly added in this curriculum revision and what content is sifted. In general, both primary and junior high school teachers welcomed these new changes, finding them very helpful to understand the composition of content.

3.2 Pedagogy/Teaching and Other Innovations That Are Prevalent or Trending

We can observe some trends in pedagogy/teaching for primary science in Japan in the national curriculum. As described in the overall objectives in the COSES, problem-solving abilities are emphasized in the primary science curriculum. Problem-solving in the context of Japanese primary science often refers to a series of steps of scientific inquiry. One of the commonly understood processes of problem-solving consists of "contact with a natural phenomenon," "awareness/questions," "understanding the questions," "assumption/hypothesis," "experiment design," "carrying out the experiment," "obtaining results," "discussion," "conclusion," and "delivery" (MEXT 2006). *Shougakkou gakushuu sidou youryou kaisetsu rika-hen* [Explanations for the Course of Study for Elementary Schools: Science] specifies that there are important problem-solving abilities for each grade, such as comparing phenomena (Grade 3), relating a change with its contributing factors (Grade 4), controlling conditions when observing/experimenting (Grade 5), and reasoning (Grade 6) (MEXT 2008b). Many teachers and researchers recognize the abilities at each grade, and they are quite happy to see the way these abilities are highlighted (Goto and SJST 2015). Some subject matter or content is intentionally arranged in such a way that they are well-related to the competencies, and this would thus show how the particular competencies in science are useful and meaningful when solving the given problems (Matsubara 2015).

Although the primary science curriculum at the national level is quite influential as previously discussed, it is also commonly accepted and valued that teachers' learning community (Inagaki and Sato 1996; Sato 2006) contributes to innovations in science teaching and learning in Japan. Examples of these teacher contributions can be found in papers/articles collected in the *Science Education Monthly*, which is one of the journals of Society of Japan Science Teaching (SJST)[1]. The authors of the papers/articles of *Science Education Monthly* consist largely of teachers of ele-

[1] The other journal of SJST is *Journal of Research in Science Education*, and this is more academic one.

mentary schools, junior high schools, and high schools as well as researchers in science education in Japan. Table 3.2 shows samples of translated titles of papers/articles of local innovations written by elementary school teachers taken from recent editions of the *Science Education Monthly*.

As shown in Table 3.2, elementary school teachers in Japan study practices of science lessons, contributing to innovations in science teaching and learning. Some of these teachers publish books that include their new teaching and learning materials often with lesson plans. These books can be purchased and used by newly employed elementary school teachers as well as science-orientated elementary school teachers. Together with the prescribed direction from the primary science curriculum at the national level, this kind of collaborative learning among teachers is believed to maintain and improve the overall level of science teaching and learning in Japan.

3.3 Assessment and Testing Practices

3.3.1 Evaluation of Learning Status as per Viewpoint (Criterion)

There are no entrance examinations for public junior high schools in Japan. It means that elementary school pupils who intend to enter public junior high schools do not have to cram for entrance examination, unlike the situation for the entry to universities. This would enable elementary school teachers more room to focus on seeing change and improvement of individual pupils. To capture change in learning of individual pupils, "*Kantenbetsuhyouka* or Evaluation of Learning Status as per Viewpoint (criterion)" and "*Hyoutei* or Rating" are used to record pupils' learning at the elementary schools in Japan.

According to the ministry's report on evaluation of pupils' learning (MEXT 2010), Evaluation of Learning Status as per Viewpoint (criterion) and Rating are based on "criterion-referenced evaluation (absolute evaluation)." When analyzing the learning status or accomplishments of individual pupils, learning contents of each subject are categorized in several criteria in the reflection of the aims described in the Course of Study. Evaluation of Learning Status as per Viewpoint (criterion) evaluates the pupil's learning attainment and clarifies the degree of achievement on a prepared set of standards. During this process, teachers often use a variety of resources that include pupils' responses and attitudes in lessons, pupils' descriptions for their homework and handouts, and results of small tests and the end-of-term examinations.

On the other hand, "Rating" looks at the overall profile of learning. Together, these two instruments aim to improve teaching and learning and ensure that individual pupils have a good understanding about the subject matter. The information

Table 3.2 Papers/articles of local innovations by elementary school teachers

Issue	Feature of the edition	Title/author/affiliation
2015/11/ Vol.64/ No.760	Active Learning in the Science Classroom	Learning Environment Design for Active Learning: Active and Collaborative Science Lessons in 3rd Grade
		Toshiaki Nishida, Idogaya Elementary School
		How to Teach Science Incorporating Active Learning into Classrooms: Learning Design Using Jigsaw Technique
		Hiromitsu Kakinuma, Sukage Elementary School
		A Better Way of Evaluating Students' Observation of Plants and the Utilization in Life & Earth Science: "Seasons and Living Things" in 4th Grade
		Hideaki Teshirogi, Nishihara Elementary School
		Support that Actively Involves Students in Problem-Solving, Tailored to Individual Needs
		Masae Yoshioka, Nakanoshima Elementary School
2015/12/ Vol.64/ No.761	This Very Moment, This Event that Changed My Lesson	What "Heart" Teaches Me: The Practice of "Structure and Functions of Human Body" in 6th Grade
		Shintaro Sato, Kita Elementary School
		Both Teachers and Children Should Develop Problem-Solving Skills! "Burning Substances in Air" in 6th Grade
		Hitomi Otani, Sunamachi Elementary School
2016/02/ Vol.65/ No.763	Expectations for the Next Curriculum Guidelines	The Learning in which Children Find that Adachi City Is Flat Lowlands: Educational Practice of "Composition of the Strata and Their Changes" in 6th Grade
		Hideaki Teshirogi, Nishihara Elementary School
		Project-Based Learning for Dealing with Problem-Solving, with Children Not Just Sit Back and Watch It Happen: Educational Practice of "Burning Substances in Air" at 6th Grade
		Hirotaka Fujisaki, Elementary School Attached to Kagoshima University
2016/03/ Vol.65/ No.764	To Learn from Classroom Research (Lesson Study)	What We Can Learn from Classroom Research Workshop
		Kenji Tanaka, Seta Elementary School
		Aiming for a Reform in Science Teaching: Our Efforts to Assign Science Coordinator to School Duties
		Shingo Imai, Hayadori Elementary School
		To Enhance the Ability to Give Lessons, Centering on Formative Assessment
		Toshiaki Nishida, Idogaya Elementary School

Source: Science Education Monthly Nos. 760, 761, 763, and 764 prepared by SJST. Retrieved from http://www.sjst.jp/magazine/ (in Japanese)

from these evaluations are shared with their parents/guardians at the end of term with report cards called "*Tuuchihyou*."

While the report cards are used for sharing, "*Shidou youryoku* or Cumulative Record of Education" is used for recording. Cumulative Record of Education, which follows a legal requirement in Japan, retains the information from "Evaluation of Learning Status per Viewpoint" and "Rating" in the section of learning records by each subject. It also records pupils' school register information, such as attendance.

In order to support teachers and schools for evaluating pupils' learning, the Curriculum Development Center in National Institute for Educational Policy (NIER) provided the "Research and Development for Evaluation Standards and Methods" in 2011. This deals with reference materials for development of evaluation standards and improvement of evaluation methods. It shows examples of standards for Evaluation of Learning Status as per Viewpoint (criterion). These examples are however regarded as *reference materials* since MEXT as well as NIER respects the prerogative of local schools when evaluating pupils' learning. Teachers should assume primary responsibility and role for evaluating pupils at the elementary schools.

3.3.2 Evaluation of Learning Status per Viewpoint (Criterion) for Science

In primary science, there are four criteria of evaluation shown below, and they are evaluated in three point-scale (1.Interest, motivation, and attitude toward natural events/phenomena is evaluated in two point-scale in some cases).

1. Interest, motivation, and attitude toward natural events/phenomena
2. Scientific thinking and expression
3. Observational and experimental skills
4. Knowledge and understanding about natural events/phenomena

Three evaluation levels are applied when recording the Cumulative Record of Education, with A being "fully satisfactory," B being "mostly satisfactory," and C being "requires efforts." Table 3.3 shows four viewpoints and their intents (aims). Teachers create an evaluation system based on these four viewpoints for each course unit. These four viewpoints are explained below.

The evaluation of "expression" should not only be based on organized tables and figures. It is expected to be based on whether or not pupils are able to logically explain their thought process, using problem-solving abilities, such as "comparing facts/events," "relating facts/events," "being aware of relevant conditions," "making deductions," and other relevant expression.

For the viewpoint of "observational and experimental skills," pupils are evaluated based on whether or not they are able to appropriately conduct observations and experiments with an awareness of the purpose of the activity.

Table 3.3 Four criteria and their intents (descriptors) for primary science

Interest, motivation, and attitude toward natural events/ phenomena	Scientific thinking and expression	Observational and experimental skills	Knowledge and understanding about natural events/ phenomena
To become familiar with nature, to engage in activities for studying natural events and phenomena with motivation, and to love and utilize nature in everyday life	To resolve problems through observing and expressing results obtained by identifying problems from natural events/phenomena, comparing events with their own prospectus, relating different events, being aware of relevant conditions, reasoning the contributing factors, and studying relevant topics	To be able to observe natural events/ phenomena, systematically conduct experiments, effectively use equipment and machines for their appropriate purpose, and accurately record the process and results	To have a realistic understanding of the characteristics, regularity, and relationships of natural events/ phenomena

Source: Curriculum Development Center (2011)

These evaluation standards are more specified for each grade level, the grouping such as “A. Matter/Energy” and “B. Life/the Earth” and course unit. As examples, four criteria and their intents (descriptors) for the grouping A. Matter/Energy is shown in Table 3.4. The evaluation standards here are based on reference material from the Curriculum Development Center.

In each school, an evaluation standard must be created upon understanding the local situation of the region and the school; however, in reality, it is assumed that many schools use the reference materials for the evaluation standard for each grade and make adjustments as needed for course units (Takagi 2011). Evaluations of four viewpoints in each course unit are used as informational material for Evaluation of Learning Status as per Viewpoint (criterion) to be recorded on Cumulative Record of Education. In the past, however, the Ministry of Education (MOE) has carried out several national level assessments as I describe next.

3.3.3 *National Assessment of Academic Ability*

For some time, there was a period of absence of a national assessment targeting pupils in compulsory education in Japan. Between 1961 and 1964, MOE implemented the national tests called *Gakute*, which targeted all Grade 8 and 9 pupils in junior high schools. The tested subjects included Japanese, mathematics, social studies, science, and English, and the results were analyzed by individual, by school, and by prefecture. There were criticisms against the national tests as some people thought that it led to unnecessary school competition and made lessons rigid.

Table 3.4 An example of four criteria and their intents (descriptors) for the grouping A. Matter/Energy

Interest, motivation, and attitude toward natural events/phenomena	Scientific thinking and expression	Observational and experimental skills	Knowledge and understanding about natural events/phenomena
To have interest in phenomena caused by the combustion of plants and try to examine the mechanism of burning on their own accord	To predict, hypothesize, probe, express, and reason the mechanism of combustion while making connections between combustion and a change in the air	To devise ways to examine combustion of plants and safely conduct an experiment with the appropriate use of tools such as gas detection tubes and a solution of calcium hydroxide	To have an understanding that during the combustion of plants, oxygen in the air is used up and carbon dioxide is produced
To take another look at phenomena around them by applying the mechanism of combustion	To use the results of self-conducted experiment, predictions, and hypothesis to reason and express opinions about combustion and a change in the air	To examine the combustion of plants and characteristics of the air and record the process and results	

Source: Curriculum Development Center (2011)

In 2007, MEXT introduced the National Assessment of Academic Ability (NAAA) targeting for all 6th and 9th graders to maintain and enhance equal opportunity in compulsory education. Under the period of the Democratic Party of Japan, sampled 6th and 9th graders participated in NAAA in 2010 and 2012, while it was not carried out in 2011 due to a national crisis. The NAAA collects data of pupils' academic abilities and background data at the national level by using assessments and questionnaires. Japanese language and mathematics are assessed, and pupil questionnaire and school questionnaire are also carried out. The pupil questionnaire asks about pupils' learning conditions and environment, such as everyday schedule, self-esteem, family relationship, reading habit, attitudes toward learning and assessed subject, etc. The school questionnaire asks about class size, teachers and pupils in the school, activities to improve academic abilities, school and curriculum management, used pedagogies and disciplines, assessment, use of ICT, utilization of the result of NAAA, special needs education, use of local resource, professional development opportunities, etc.

The Assessment of Science is included every 3 years which started in 2012. Lasting 1 day, the NAAA is implemented on the same day nationwide in April, which is the first month of the academic year in Japan. Later, the individual results are returned to each pupil. In order to avoid unnecessary competition among schools, MEXT does not release information or results that show performance of individual schools, while the results of analysis from the prefectures are released. The scores of each prefecture have been drawn keen attention from the public. While in the relatively higher-performing prefectures, their good practices are shared through a number of educators visited, in the relatively lower-performing prefectures, considerations are made aiming to improve the teaching and learning, resulting in more investment in education, for example.

According to MEXT, "The results have revealed that there have been challenges in the utilization of knowledge and skills. Educational policies and classroom teaching have been improved based on these results" (cited from *http://www.mext.go.jp/english/elsec/1303755.htm*). As previously mentioned, one of the aims of NAAA is to maintain and enhance equal opportunity in compulsory education as well as to ensure the quality. Many of the efforts and initiatives related to educational reforms made at each prefecture and board of education are now based on or use the results and practical data provided by NAAA.

Similar to PISA's test items, NAAA's test items have an emphasis on providing a context to the test items. By doing so, NAAA tries to show a relationship between content of the discipline and real world. In the case of science, real-world context is often given to stimulate students' questions to a natural phenomenon (item (1) in Question 3 below), and this is expected to lead students to form their predictions. Some test items, such as item (2) in Question 3 below, are designed to ask if students can plan a rational test or experiment to verify a prediction. This is one of the ways that NAAA science test items try to measure students' scientific thinking and expression.

Unlike PISA, every student in the target grade attempts the NAAA. This means that teachers in every school in Japan have opportunity to actually see the test items on paper. So, naturally, teachers discuss the items and at times talk about whether they are good ones or might be different from the items they usually make for their end-of-term tests. Some teachers might point out that NAAA's test items emphasize measuring scientific thinking and expression in a real-world context rather than checking factual knowledge only. The presence of NAAA's test items is now contributing the discussion of classroom teaching and learning in science.

The box below shows items (1) and (2) in Question 3 of Elementary School Science of the 2015 National Assessment of Academic Ability (CDC 2015a) (translated by author).

Question 3

Yukari and her friends decided to make iced tea with milk and to make a sugar solution to put into it.

(1) To make the iced tea with milk, Yukari and her friends put cold water into a jug and boiled it.

Choose one answer from options 1 to 4 below corresponding to the missing words (A) in Toshio's statement and write down the number of the answer you have chosen.

1 Turned into a gas and become invisible
2 Turned into a liquid and become invisible
3 Turned into a solid and become invisible
4 Disappeared and become invisible

(2) When they put tea leaves into the boiling water, the tea leaves moved around inside the jug.

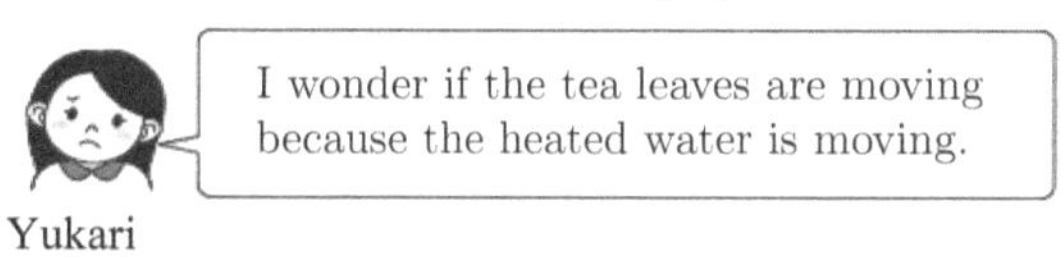

Based on the fact that the tea leaves were moving, Yukari and her friends posed the question “How does water heat up?” and drew diagrams showing their predictions for what would happen.

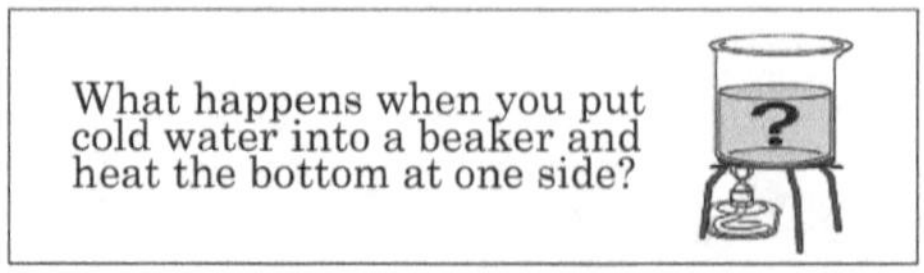

To find out whether their predictions were right, Yukari and her friends decided to carry out an experiment, putting three thermometers – A, B, and C – into the beaker, as shown in the diagram below.

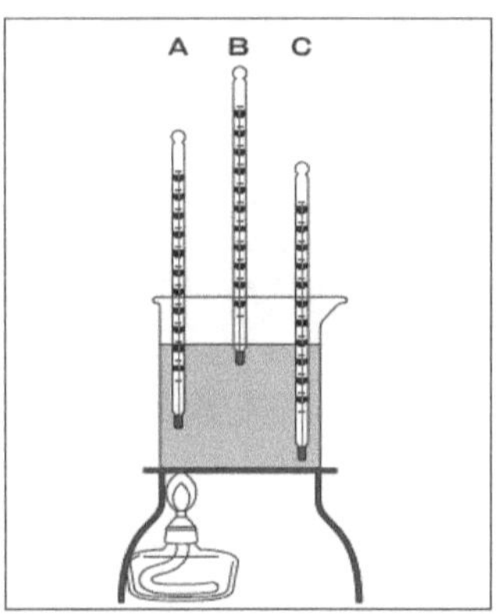

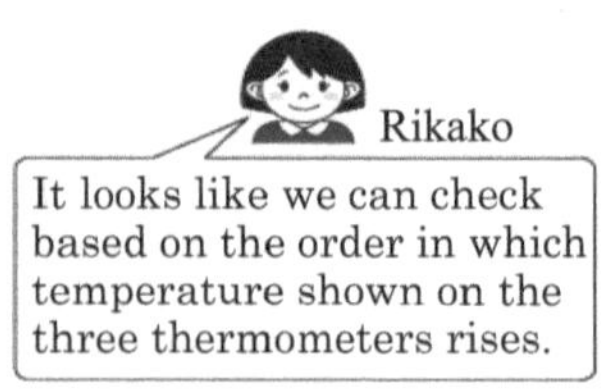

If Rikako’s prediction is right, what is the order in which the temperature shown on the thermometers will rise? Choose one answer from options 1 to 4 below and write down the number of the answer you have chosen.

1 B→A→C

2 B→C→A

3 C→A→B

4 C→B→A

The box below shows the explanations about items (1) and (2) in Question 3 of Elementary School Science of the 2015 National Assessment of Academic Ability described in the document called *Explanation materials for the 2015 National Assessment of Academic Ability: Elementary School Science*. These materials were prepared by the Curriculum Development Center National Institute for Educational Policy Research (CDC 2015b) (translated by author)

1. Purpose of Setting the Question

This question examines whether or not students can plan an experiment by predicting the results that will be forthcoming if their expectation is correct, and whether or not they can improve their ideas based on the results of the experiment, in order to examine how water heats up.

Focusing on the movement of tea leaves, it asks how water heats up.

Pupils are required to plan an experiment after first predicting what result from the experiment would prove their expectation about the answer to the question to be correct. In addition, after the experiment, they are required to use the results obtained from the experiment to judge whether or not their expectation was correct, and, if necessary, to make improvements to develop a more reasonable idea. When conducting the experiment, it is necessary to formulate a proper prediction or hypothesis, and to approach the experiment with this prediction in mind.

2. Commentary

Question (2) Purpose

This question examines whether or not students can plan an experiment by predicting the results that will be forthcoming if their expectation is correct.

■ **Grouping and Content described in the Course of Study**

[4th Grade] A Matter/Energy

(2) To develop pupils' ideas about the properties of metals, water and air, by exploring the changes in warming and cooling metals, water and air.

(b) The temperature of metals goes up gradually, spreading from the point being heated, but heated air and water move, raising the temperature of the whole.

Objective of Evaluation: Scientific thinking and expression

Framework: Mainly questions about application (planning)

To investigate how the whole volume of water heats up as the portion in contact with the heat moves, it is important to predict how the heated water will move and to plan an experiment based on a prediction of the order in which the temperature indicated on the thermometers placed in the beaker will rise. Here, if the water heats up in the way predicted by Rikako, the water around the bulb of thermometer C should heat up first, then the water around the bulb of thermometer B, and then the water around the bulb of thermometer A, so it is important to choose option 4.

3.4 Recent Reform Attempts

3.4.1 Background of the Revision

Under Japan's curricular reform, the national curricula or the Courses of Study have been revised eight times since their implementation in 1947 as previously mentioned. These revisions were done with the goal of keeping up with societal changes over the years and the needs of each age group. Like other countries, preparing the young for upcoming challenges in new era has been one of the important aspects to consider in the curriculum revision in Japan. On 28th of January 2005, the Central Council for Education released the report "Future Vision of Higher Education in Our Country," stating that the twenty-first century is a "knowledge-based society." Some of the characteristics of this society described in the report are (1) knowledge has no borders, and globalization advances even more; (2) knowledge rapidly progresses every day, and there are constant competition and innovation in technology; (3) advancement in knowledge often comes with a change of the paradigm, making it more important to be able to make judgments based on extensive knowledge and flexible cognitive skills; and (4) participation regardless of gender and age is promoted (Central Council for Education 2005).

In February 2005, in order to enhance education for children in the twenty-first century, the Minister of Education, Culture, Sports, Science and Technology requested that the Central Council for Education examine revisions to the entire standards for the curriculum in Japan, as well as to improve the quality (competencies) of teachers and education conditions. The Council began deliberations in April 2005. During this period, amendments to the Basic Act on Education and to the School Education Act were made, which legally specified the necessity of harmonic development of balancing to foster an attitude to acquire wide-ranging knowledge and culture and to seek the truth and cultivate a rich sensibility and sense of morality, while developing a healthy body (Article 2 (1) in the Basic Act on Education), while focusing on basic/fundamental knowledge/skills, thinking/judging/expressing abilities, and enthusiasm for learning (Article 30 (2) in the School Education Act) in school education. After almost 3 years of deliberations in the Central Council for Education, taking into account the law revisions that stretch back to the foundation of education, the Council published the report "*Youchien, syougakkou, chuugakkou, koutougakkou oyobi tokubetsushiengakkou no gakusyuusidouryouryoutou no kaizen ni tuite (toushin)* or Report on the improvement of national curriculum in kindergartens, primary, lower secondary and upper secondary schools, and special schools" in January 2008. In this report, the directions for the revisions of the Courses of Study were shown for each school and each subject, taking into account the challenges of children today. It states that the revisions of the Courses of Study for this time should be based on the fundamental principles shown by the revised Basic Act on Education, and other important points to consider are:

1. Sharing of the philosophy, "Zest for life"
2. Acquiring basic and fundamental knowledge and skills
3. Nurturing the ability to think, decide, and express
4. Securing the necessary hours of class time to establish definite academic performance
5. Improving learning motivation and establishing learning habits
6. Improving guidance for the nurturing of rich humanity and a healthy body (Source: Central Council for Education 2008)

These important points to consider have become fundamental for revisions of curriculum for each subject.

3.4.2 Influence of International Large-Scale Assessments Such as TIMSS and PISA

Japan has been participating in international large-scale assessments such as TIMSS (Trend in International Mathematics and Science Study) and PISA (Programme for International Student Assessment) since their emergence decades ago. There have been some influences from the assessments on the revision of primary science curriculum. However, before jumping into actual influences caused by those assessments, it is beneficial to look at the differences of TIMSS and PISA as the types of information the two assessments provide determine types of influences.

According to the TIMSS 2015 Assessment Framework (Mullis and Martin 2013), it is organized around two dimensions: one is the content dimension, specifying the subject matter (life science, physical science, earth science), and the other is the cognitive dimension, specifying the thinking processes (knowing, applying, reasoning). It is rational to think that the results of each items of TIMSS assessment are related to students' actual learning at schools as well as the curriculum coverage including its emphasis at the national level (Matsubara and Hagiwara 2014). On the other hand, PISA 2015 framework focuses on scientific literacy that has four components such as competencies, knowledge, attitude, and context (OECD 2013). Many of the actual items in PISA try to measure competencies in certain contexts, and they are not so directly related to school science content. We can say that TIMSS is a type of assessment that is more curriculum based compared to PISA. Naturally, then, TIMSS's results have more detailed information about curriculum revision. Accordingly, some Japanese researchers in science education did empirical studies by using data taken from past TIMSS science assessments (e.g., Saruta 2007; Hagiwara and Matsubara 2014; Matsubara et al. 2016). These studies tried to contribute to curriculum revision providing information about curriculum or relationship between curriculum and achievement on a topic- or item-based analysis of Japan's and/or other countries'. However, the influences of these studies are rather indirect. The contributions from these studies have been intended to support discussion of curriculum revision held among curriculum experts and science teachers.

PISA is domestically more widely recognized. Many think the assessment can work to measure competencies that are needed to live well after schooling or compete globally. Although PISA's results themselves do not tell much about what is in particular needed for a country's curriculum, for example, which topics are more important for a particular grade of a country's curriculum, it is thought to be more influential in general. According to Nakayasu (2016), the Japanese government has placed importance on the cultivation of human resources for international competitions, and the PISA has had a large influence on the direction of Japanese education. When it comes to science education, PISA's items have stimulated teachers and researchers who value science processes and connection to real world. Now, test items, which are trying to measure scientific competencies in given contexts, are often called PISA *gata mondai* or PISA-type problems.

PISA's influence can be seen in an official document, too. When explaining one of the background information for the curriculum revisions in 2008, MEXT (2008b) pointed out several challenges Japanese students faced, mentioning that the challenges found were based on various surveys such as the PISA. The followings are the challenges described in MEXT (2008b):

1. Challenges in answering reading literacy questions and essay-type questions, which require thinking/judging/expressing abilities, as well as questions that require utilization of knowledge and technologies
2. Challenges in enthusiasm for learning, learning habits, and lifestyles, including study time at home, which were reflected in the increasingly dispersed distribution of the reading literacy test results
3. Challenges such as lacking self-confidence, anxiety about the future, and a decrease in physical fitness.

It can be noted that challenges (2) and (3) are related to attitudinal challenges for learning. The descriptions are based on results of the questionnaires administered as a part of PISA and also TIMSS. The international large-scale assessments have had an influence on this kind of non-cognitive aspects of learning, too.

3.4.3 Revisions of Primary Science Curriculum

Together with the important points explained above, the report (Central Council for Education 2008), specified the important areas that should be improved through this revision in each subject. The revision of the primary science curriculum was based on the report, like the revisions of other subjects. Focuses of the improvements are enabling pupils to find problems in a familiar natural environment by themselves based on learning content of living environment studies, developing problem-solving abilities through carrying out observations and experiments with their own prospectus, developing a realistic understanding of the learning content in relation to real life, fostering an attitude of respect for the natural environment and life and to scientifically explore, and developing scientific perspectives and ideas.

For improvement of the content of elementary school science, the following contents are added/transferred/deleted/transferred and integrated into lower secondary school education. This reorganization was made trying to contribute to developing the abilities of problem-solving, fostering an attitude for loving nature, developing a realistic understanding of natural phenomena, and developing scientific perspectives and ideas. The box below shows the added and transferred content for the revision of the primary science curriculum in 2008.

	Grade 3	Grade 4	Grade 5	Grade 6
Added content	Object and weight	Changes in water volume	Small living things in water	Use of a lever
	Function of wind and force of rubber	Structure and movement of the human body	Upstream, downstream, and pebbles on a riverside	Use of electricity
	Observation of familiar environment		Cloud and weather change	Presence of major organs
				Pathway of water
				Relationship of living things through food
				The moon and the sun
Content transferred		The change in temperature in a day (moved from Grade 5)	Function of electricity (moved from Grade 6)	Regularity of a lever (moved from Grade 5)

Content transferred and integrated to lower secondary school course
Collision of objects (previously Grade 5)

In summary, the following are some of the major revisions for the primary science curriculum.

1. In "Matter/Energy" and "Life/the Earth," instruction is focused on exploration using observation, experiments, and models as well as making, and the content is structured accordingly.
2. The basic perspectives and concepts in science, such as "energy," "particles," "life," and "the earth," are used to systematize the learning contents.

3. Importance is placed on learning activities of organizing, examining, and expressing the observational and experimental results in order to improve the pupils' scientific perspectives and ways of thinking.

These curricular revisions were made for the children who live in knowledge-based society in this twenty-first century to develop basic knowledge and skills, understanding of scientific concepts, and the ability to think, judge well, and express their opinions and ideas to others using those scientific knowledge and concepts.

3.5 Local Innovations

The "Period for Integrated Studies" is one of the characteristics of Japanese curricula. It allows schools and teachers to develop school curricula that meet needs of individuals in their specific localities. Each school can determine the course title for the Period for Integrated Studies. Table 3.5 shows the yearly distribution of lesson hours for the Period for Integrated Studies and science.

The "Period for Integrated Studies" was discussed in the Course of Study for the first time in the 1998 revision of the Course of Study. In the 2008 revision of the Course of Study, it is discussed in a whole new chapter. The objective for Period for Integrated Studies is shown below.

> To enable pupils to think in their own way about life through cross-synthetic studies and inquiry studies, while fostering the qualities and abilities needed to find their own tasks, to learn and think on their own, to make proactive decisions, and to solve problems better. At the same time have them acquire the habits of studying and thinking, cultivating their commitment to problem solving and inquiry activities in a proactive, creative and cooperative manner. (MEXT 2008a)

Table 3.5 Yearly distribution of lesson hours for Period for Integrated Studies and science

	Number of hours for Period for Integrated Studies	Number of hours for science	Total number of hours for the grade
Grade 3	70	90	945
Grade 4	70	105	980
Grade 5	70	105	980
Grade 6	70	105	980

Source: Ordinance for Enforcement of the School Education Act Appendix 1

As described in the objective, the concepts of "fostering the problem-solving qualities and abilities" and "to think in their own way about life" enable pupils to achieve "learning how to learn." The unique characteristic of "Period for Integrated Studies" in the 2008 revision is the addition of "through inquiry studies" in the first part of the objective (Matsumoto 2011).

The section of "Syllabus Design and Handling the Content" in Chapter 5 "The Period for Integrated Studies" of the Course of Study describes several points that should be considered in designing the syllabus. The box bellow shows some of the points described in the section.

(1) The overall plan and annual teaching program should include objectives and content, pupils' qualities, abilities and attitudes to be developed, learning activities, teaching methods, instructional systems, and plan for evaluating learning outcomes while ensuring a relationship with all educational activities of the school
(2) Taking into account the circumstances of the local community, school, and pupils, each school should conduct educational activities that abound in ingenuity such as cross-synthetic studies that cross over the boundary of subjects and periods, inquiry studies, and studies based on pupils' interests and concerns
(3) With respect to objectives and content to be established by each school listed in Subsection II, relationships with everyday life and society should be emphasized
(4) Pupils' qualities, abilities, and attitudes to be developed should be based on the viewpoints of matters related to learning methods, pupils themselves, and the involvement of others and society
(5) Learning activities should be conducted in the light of the conditions of each school, for example, learning activities about cross-synthetic tasks, including international understanding, information, environment, and welfare/health, learning activities about tasks based on pupils' interests and concerns, and learning activities about tasks depending on characteristics of the local community and the school such as people's lives, traditions, and culture of the local community
(6) Teachers should make an effort to connect knowledge and skills acquired in each subject, moral education, foreign language activities, and special activities with each other and to utilize them in learning and living so that such knowledge and skills can be applied comprehensively.

Source: MEXT (2008a), Chapter 5 The Period for Integrated Studies, III. Syllabus Design and Handling the Content (extracted, English translation is based on http://www.mext.go.jp/component/a_menu/education/micro_detail/__icsFiles/afieldfile/2011/10/12/1261037_13_1.pdf

Point (1) in the box discusses "Period for Integrated Study" needs to be related to the whole activities of the school. Point (2) emphasizes educational activities need to be well-thought and planned so that they can be interdisciplinary or inquiry activities, meeting pupils' interests and concerns. From these descriptions of point (3) to (5), it is seen that connecting school education to locality is one of the aspects of "Period for Integrated Studies." In addition, point (6) talks about integration of knowledge which is also one of the important aspects of this initiative. In these ways, several important points for developing school curricular are explicitly given, and this contributes to supporting the local innovations at the school level.

In order to help the local innovations in a more concrete way, the ministry collects and provides examples of good practices on its web. Table 3.6 shows examples

Table 3.6 Examples of good practices of "Period for Integrated Studies" developed by local elementary schools (translated by author)

Title of "Period for Integrated Studies"	School
"Dream Time" for finding a problem, finding myself, reaching out our community	Sapporo city municipal Shiseikan Elementary School
"Kashiwa Learning" for fostering ability to learn, ability to relate, and ability to improve	Akita city municipal Kawashiri Elementary School
"Ooka Period" for fostering nine competencies	Yokohama city municipal Ooka Elementary School
"Takatomi Pupils' Time" for fostering pupils loving Takatomi and deepening inquiries independently	Yamagata city municipal Takatomi Elementary School
"Inasa Challenge Time" for fostering ability for problem-solving and ability for relating with people appropriately	Nagasaki city municipal Inasa Elementary School

Source: http://www.mext.go.jp/a_menu/shotou/sougou/1300434.htm

of good practices of "Period for Integrated Studies" developed by local elementary schools, which can be found in the ministry's web. It is emphasized that these good practices remain examples because the ministry intentionally stays on the side of not disturbing schools' initiatives.

Akita city municipal Kawashiri Elementary School, for instance, has developed a local curriculum for the Period for Integrated Studies whose title is called "Kashiwa Learning" for fostering ability to learn, ability to relate, and ability to improve. The goal of their initiative is to foster students' abilities to learn using inquiry and create their own learning, through providing the learning based on daily lives and experiences. They also try to utilize local human and physical resources, which can help students actively relate with others and use what they have learnt. Their learning objectives and content have been developed considering developmental stages of students and relationship with disciplines. In order to develop units that are suitable with the actual situations of the students, the learning objectives and content were intentionally made not too be specific. Each grade has keywords or themes of the learning content, such as Locality for Grade 3, Environment for Grade 4, Life for Grade 5, and Human for Grade 6. Then, specific learning content will be decided at the beginning of the school year after considering students' actual needs.

Responding to the keyword for Grade 3, which is Locality, the learning objectives in the school year of 2010 were to deepen understanding about our locality and think about good aspects of it. The learning content for the objectives was local facilities and natural resources and people keeping local heritage and culture. In Grade 4, as the keyword is Environment, deepening understanding about environmental issues and thinking about what we can have become the learning objectives of Grade 4 in the school year of 2010. All other specific details of the learning content are described in the whole plan of "Kashiwa Learning," which is prepared annually by the school. Table 3.7 shows aspects of learning activities, instructions, systems, and assessment of "Kashiwa Learning" that are extracted from the whole plan in the school year of 2010.

Table 3.7 Some aspects of "Kashiwa Learning" in the school year of 2010 (extracted)

Learning activities	Units are developed in each grade group, and activities can be modified depending on development of the lessons
	Each grade group makes one theme a year
Instructions	Pay attention to continuities of inquiry learning and promote mutual learning
	Try to relate well with other disciplines
Systems	Establish instruction and support systems among all the teachers, and try to communicate well
	Use local resources such as *local home teacher*
Assessment	Put more emphasis on individual assessment, and use the result for instruction
	Enrich assessment by using portfolio.

Source: http://www.mext.go.jp/component/a_menu/education/detail/__icsFiles/afieldfile/2011/02/17/1300459_11.pdf

The "Period for Integrated Studies" is certainly one of the features of Japanese curricula, and local innovations are expected here. It provides a meaningful learning environment through cross-disciplinary or interdisciplinary approaches as explained in the objective. Students are expected to learn and think on their own, to make proactive decisions, and to solve problems better in a cooperative manner. As schools and teachers have chances and responsibilities to develop school curricula for actual needs of individuals and localities, this initiative might be, in the future, recognized as one start of decentralization of Japanese education.

References

Central Council for Education. (2005). *Future vision of higher education in our country.* (in Japanese). Retrieved from http://www.mext.go.jp/b_menu/shingi/chukyo/chukyo0/toushin/05013101.htm

Central Council for Education. (2008). *Report on the improvement of national curriculum in kindergartens, primary, lower secondary and upper secondary schools, and special schools.* (in Japanese). Retrieved from http://www.mext.go.jp/a_menu/shotou/new-cs/news/20080117.pdf

Curriculum Development Center, National Institute for Educational Policy Research. (2011). *Research and development for evaluation standards and methods.* (in Japanese). Retrieved from http://www.nier.go.jp/kaihatsu/hyouka/shou/04_sho_rika.pdf?time=1471415939179

Curriculum Development Center, National Institute for Educational Policy Research. (2015a). *Assessment items for the 2015 national assessment of academic ability: Elementary school science* (in Japanese). Retrieved from http://www.nier.go.jp/15chousa/pdf/15mondai_shou_rika.pdf

Curriculum Development Center, National Institute for Educational Policy Research. (2015b). *Explanation materials for the 2015 national assessment of academic ability: Elementary school science* (in Japanese) Retrieved from http://www.nier.go.jp/15chousa/pdf/15kaisetsu_shou_rika.pdf

Goto, K., & Curriculum Committee of SJST. (2015). Perspectives of the members of SJST towards competencies emphasized in the Course of Study for science. *Proceedings of annual conference of Society of Japan Science Teaching, 105* (in Japanese).

Hagiwara, Y., & Matsubara, K. (2013, June). A DIF analysis of TIMSS–2007 assessment in physics and chemistry focusing on the matching of the test items and the curricula: The comparison of Japanese and Korean eighth graders. *Paper presented at the 5th IEA International Research Conference*, Singapore. Retrieved April 14, 2015, from http://www.iea.nl/fileadmin/user_upload/IRC/IRC_2013/Papers/IRC-2013_Hagiwara_Matsubara.pdf.

Hagiwara, Y., & Matsubara, K. (2014). An IRT analysis of TIMSS–2011 grade eight test items in physics and chemistry, in consideration of test-curriculum matching: A comparison of Australia, England and Singapore. *Japanese Journal for Research on Testing, 10*, 95–113 (in Japanese).

Inagaki, T., & Sato, M. (1996). *An introduction to lesson study.* Tokyo (in Japanese).

Matsubara, K., & Hagiwara, Y. (2014). An analysis of the relationship between Japan's science achievement in international studies and changes in national curriculum: Based on the statistical changes over time in common items in recent TIMSS assessments in eighth grade biology. *Journal of Research in Science Education, 55*, 69–80 (in Japanese).

Matsubara, K. (2015). Relationship between competencies and learning content in science: Creating scenes in lesson to enable pupils to realize the necessity of competencies. Elementary School Science Education. *Nihon Syotourika Kyouiku Kenkyukai (Japan Elementary Science Education Association), 45*(3), 7–10 (in Japanese).

Matsubara, K., Hagiwara, Y., & Saruta, Y. (2016). A statistical analysis of the characteristics of the intended curriculum for Japanese primary science and its relationship to the attained curriculum. *Large-scale Assessments in Education, 4*, 13. https://doi.org/10.1186/s40536-016-0028-0

Matsumoto, S. (2011). Period for integrated studies and science education. In S. Kadoya (Ed.), *Science for new learning and horizons: Theories and practices of lessons for primary schools* (pp. 169–174). Kyoto: Minerva (in Japanese).

Ministry of Education, Culture, Sports, Science & Technology. (2006). *Images of revising learning content of science (draft)*, Central Council for Education, Primary and Secondary Education Sectional Committee, Curriculum Subcommittee (fourth stage, 11th) conference minutes/documents distributed 4–2 (in Japanese).

Ministry of Education, Culture, Sports, Science, and Technology. (2008a). *Syougakkou gakushuu sidou youryou* [Course of study for elementary school], Tokyo shoseki (in Japanese). English translation (tentative version) is available on the ministry's website. Retrieved from http://www.mext.go.jp/component/a_menu/education/micro_detail/__icsFiles/afieldfile/2009/04/21/1261037_5.pdf

Ministry of Education, Culture, Sports, Science, and Technology. (2008b). *Syougakkou gakushuu sidou youryou kaisetsu rika-hen* [Explanations for the course of study for elementary school: Science], Dainippon tosho (in Japanese).

Ministry of Education, Culture, Sports, Science & Technology. (2010). *Report on evaluation of pupils' learning* (in Japanese). Retrieved from http://www.mext.go.jp/b_menu/shingi/chukyo/chukyo3/004/gaiyou/1292163.htm

Ministry of Education, Culture, Sports, Science, and Technology. (2014a). *Gakkou kihon chousa* [Basic survey on schools]. Tokyo: Author. (in Japanese).

Ministry of Education, Culture, Sports, Science, and Technology. (2014b). *White paper on education, culture, sports, science, and technology*. Tokyo: Author (in Japanese).

Mullis, I. V. S., & Martin, M. O. (Eds.). (2013). *TIMSS 2015 Assessment Frameworks*. Chestnut Hill: TIMSS & PIRLS International Study Center, Boston College.

Nakayasu, C. (2016). School curriculum in Japan. *The Curriculum Journal, 27*(1), 134–150. https://doi.org/10.1080/09585176.2016.1144518.

Takagi, M. (2011). Recording of students' learning. In S. Kadoya (Ed.), *Science for new learning and horizons: Theories and practices of lessons for primary schools* (pp. 155–162). Kyoto: Minerva (in Japanese).

OECD. (2013). *PISA2015 draft science framework March 2013*. Retrieved from https://www.oecd.org/pisa/pisaproducts/Draft%20PISA%202015%20Science%20Framework%20.pdf

Saruta, Y. (2007). Characteristics of curriculum and scientific literacy in Japan. *NIER Research Bulletin, 137*, 27–45 (in Japanese).
Sato, M. (2006). *Challenges of schools: Creating learning communities*. Shogakukan: Tokyo (in Japanese).
Shiba, K. (2011). History of primary science education after World War Two. In S. Kadoya (Ed.), *Science for new learning and horizons: Theories and practices of lessons for primary schools* (pp. 1–9). Minerva: Kyoto (in Japanese).
Statistics Bureau and Statistical Research and Training Institute. (2016). *Nippon toukei nenkan 2016* [Japan statistical yearbook 2016]. Tokyo: Author. (in Japanese). Retrieved from http://www.stat.go.jp/english/data/nenkan/index.htm

Chapter 4
Primary Science Education in China

Weiping Hu and Xin Shou

4.1 Overview of Education in China

Education in the People's Republic of China has increasingly gained attention in recent years; the British Broadcasting Corporation's documentary "Are our kids tough enough? Chinese school" started a huge discussion about the differences in education between East and West when it was broadcast in 2015. The results from the Programme for International Student Assessment (PISA) in 2012 have also reported that students in Shanghai ranked first in mathematics, reading, and science. At the very least, it indicated that education in Shanghai was successful according to what the test had measured (http://www.oecd.org/pisa/keyfindings/pisa-2012-results.htm). The Organisation for Economic Co-operation and Development (OECD)'s latest report on "Strong Performers and Successful Reformers in Education" also suggested that Shanghai had solved a number of issues of educational equity through exemplary means. In particular, the OECD regarded the "mandatory administration" approach here as a novel policy instrument to assist weaker students in rural schools level up with their counterparts in urban schools (OECD 2013).

But education in Shanghai is certainly not representative of education in the country; what does education look like in the rest of China? The Institute of Economics in the Chinese Academy of Social Sciences predicted that growth in the national economy stood at about 6.9% in 2015. Increased investments in education reached 4.3% of GDP, which exceeded the target of 4% of GDP according to the official document "Outline of China's National Plan for Medium and Long-term Education Reform and Development (2010–2020)" (see Fig. 4.1) (Chinese Association of Science and Technology, 2002). Statistics from the Ministry of

W. Hu (✉) · X. Shou
Shaanxi Normal University, Xi'an, China

Y.-J. Lee, J. Tan (eds.), *Primary Science Education in East Asia*, Contemporary Trends and Issues in Science Education 47,
https://doi.org/10.1007/978-3-319-97167-4_4

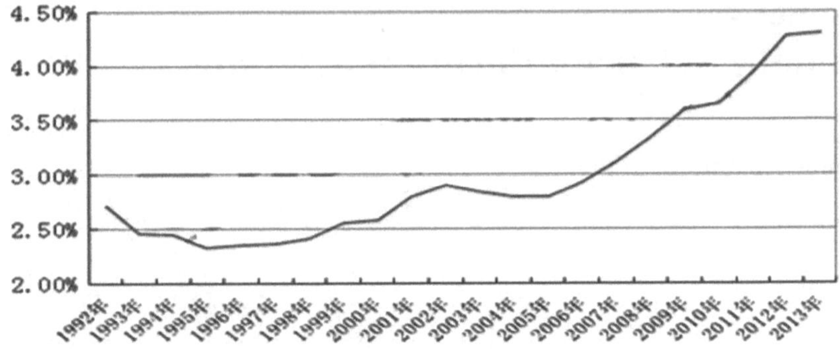

Fig. 4.1 Percentage of government appropriation for education accounted for GDP. (Data source: National Bureau of Statistics, China, 2015) from http://www.eol.cn/html/jijiao/report/2015/pc/content.html

Education further showed that net enrollment of elementary school children in 2014 stood at 99.81% with more than 61.06% of primary schools equipped with equipment for conducting experiments in science and mathematics. Gross enrollment of junior high school students was 103.5% with 95.1% of them graduating successfully. At this level, 81.33% of schools were equipped with laboratory equipment in the physical sciences. Enrollments among senior high school students was 86.5% with 87.63% of schools here having laboratory equipment in the physical sciences. Enrollment rates in higher education reached 37.5% indicating that increasing educational attainment for young people plays an important role in reducing educational inequality according to the Ministry of Education (Yang et al. 2014). Although there is still a gap between China and other developed countries, the government steadfastly holds onto two principles: to promote equality and to improve standards or quality (素质) in education. By the end of 2020, it is hoped that modernization of education and the achievement of a learning society will be realized as a first step and thus help China achieve its ambitions to improve human resources.

With the goal to achieve equity and core competencies, constant reforms in curriculum and teaching have been the norm. Recently an official report "Outline of China's National Plan for Medium and Long-term Education Reform and Development (2010–2020)" outlined in detail some of these objectives to update education in China:

- *Achieve greater standards in basic education.* By 2020, preschool education shall be universal in the country, while the implementation of 9 years of compulsory schooling (ages 7–15) shall be consolidated and enhanced. Meanwhile, high school education shall become the norm with 90% gross enrollment rates, while higher education shall be further strengthened in quality to reach 40% gross enrollment. Illiteracy shall also be eliminated among the young and the middle-aged. The average number of years of education enjoyed by new additions to the workforce shall rise from 12.4 to 13.5 years. The average number of years of education received by the working-age (20–59 years old) population shall be extended from 9.5 to 11.2 years. Twenty percent of the working-age population shall have completed higher education by 2020 thus doubling that of the 2009 figures.

- *Realize the benefits of equal opportunities in education for all.* To strengthen the benefits and rewards arising from education, education shall remain a public right, and equal access to it shall be safeguarded. All citizens should have access to good education according to the law. A basic public education service network covering both urban and rural areas shall be set up in which equal quality of services are provided and regional disparities narrowed. Stepped-up efforts should be made to run every school well and provide the best possible education to every student. No child shall leave school due to family financial woes. Equal opportunities in education for children of rural migrant workers in cities shall be provided in a realistic manner. Disabled peoples' right to education shall be also guaranteed without fail.
- *Provide high-quality education through various means.* The quality of education as a whole shall be enhanced, while its modernization shall rise by a significant margin. The total volume of high-quality educational resources shall grow steadily to better meet people's demands for quality education. Strong improvements shall be made in students' moral and intellectual development, scientific and cultural awareness, and health education. The capabilities of professionals in all fields of endeavor to serve the nation and the people and to participate in the global arena shall be greatly increased.
- *Building a comprehensive framework for lifelong education.* Diploma-granting or formal education shall develop in coordination with non-diploma-granting education. Vocational education shall be connected with regular education in schools, and there shall also be fruitful links between prevocational and on-the-job training. Participation in further or continuing education shall go up by a wide margin; as much as 50% of jobholders should receive such education. The national system of modern education shall be perfected, and a basic framework for lifelong education shall be in place, so that everyone can be taught what they want to learn, excel at what they learn, and put what they have learned into use.
- *Establishing a fully vibrant education system.* It is imperative to further emancipate our mind, update our concepts, deepen educational reforms, further increase the opportunities for education, bring about an education system that is congenial to a socialist market economy, and strive to build a middle-income society in all respects. Such an education system should also be vibrant, efficient, open, and well adapted to scientific developments and the requirements to run a world-class modern education with Chinese features.

To achieve these education goals, more efforts must be conducted in policy, curriculum, teaching, teacher training, and educational research (Adelainde 2009). Since China started major policy reforms 30 years ago, universal access to basic education and upgrading of educational standards have received serious attention by the government. In 2003, the government announced that by 2007, students in rural areas who came from families experiencing financial difficulties were entitled to enjoy the policy of "two exempt and one subsidy[1]." This policy would ensure that no student would drop out of school due to their family circumstances. In 2008, tuition and school fees in urban areas were exempted nationwide, and the children of migrant workers could receive state education in cities as part of the national system of education. However, the academic performance of the latter in migrant schools has not been as good as expected (Lai et al. 2014).

The transformatory practice of curriculum reform over the last 30 years has been guided by an initial and fundamental research question—how does it serve the people (以人为本)? "Consequently, school management philosophy has been forced to change from 'education for elites or the pursuit of success in examinations'" as major foci toward questions concerning human development, development over life, and transformations in/for life. Such reforms have been accompanied by calls for new ways of learning such as active learning, collaborative learning, and inquiry-based learning. Teaching norms are thus supposed to transform from prioritizing knowledge-intensive learning to dialogic modes, building and emphasizing the interaction between teachers and students during teaching. In this new educational context, an investigative spirit toward learning is called for among other priorities. It is worth emphasizing that changing the learning method will cause a series of profound chain reactions in the educational field, not just its effect on students. It involves changing teachers' educational beliefs and teaching, changing school leaders' management of schools, and how schools are evaluated.

In terms of teachers' professional development, teachers' specialization and lifelong learning have been general trends. As shown in Fig. 4.2, the number of primary school teachers tends to be stable, and the teacher-student ratio has fallen to 16.8 to 1 over the past 20 years (Fig. 4.3). Teaching in small class size is different from previous teacher-centered method suitable for large classes; it brings challenges at a professional level and to one's educational philosophy to create dialog-centered teaching for everyone. Now the government is formulating teacher development programs for 2020, hoping that through strong policy guidance, public financial support, and pushing for massive high-quality teacher training in the country will eventually promote teachers' professional development. The midterm goal is toward enhancing teachers' education levels significantly such that primary

[1] It means no charges of tuition and miscellaneous fees while having allowances for boarding living expenses.

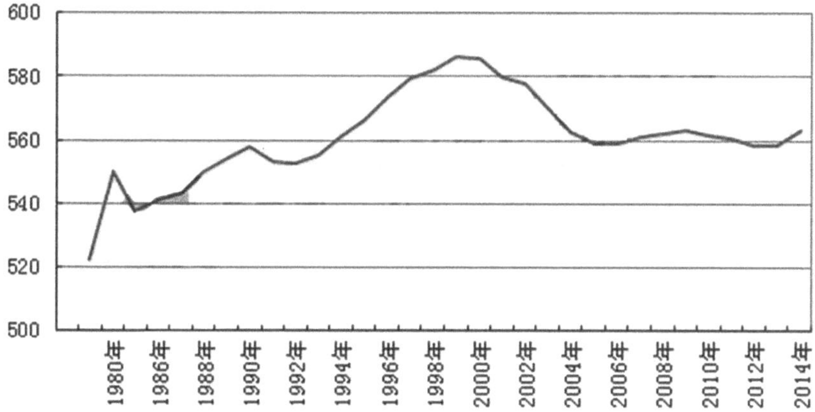

Fig. 4.2 Variation of primary teachers' quantity. (Data source: the Ministry of Education, China, 2015) from http://www.eol.cn/html/jijiao/report/2015/pc/content.html

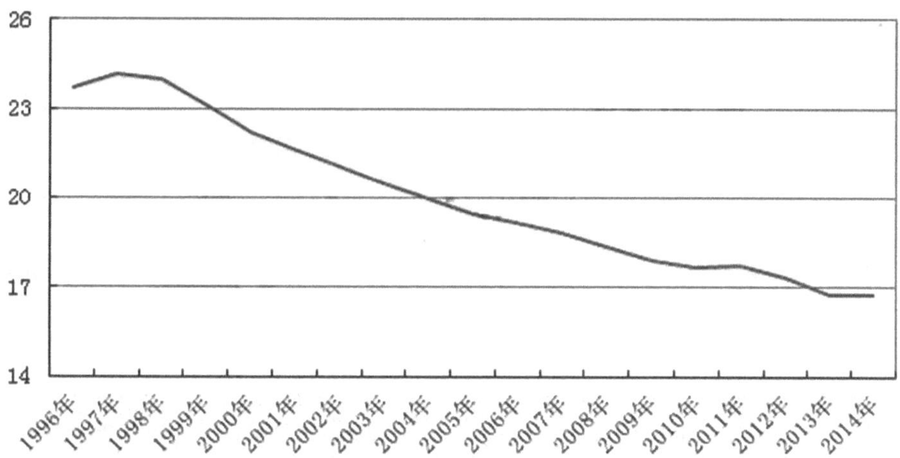

Fig. 4.3 Teacher-student ratios in primary school. (Data source: the Ministry of Education, China, 2015) from http://www.eol.cn/html/jijiao/report/2015/pc/content.html

school teachers have specialist qualifications and middle school teachers obtain bachelor's degrees. Meanwhile, comprehensive quality and teaching ability are expected to greatly improve by way of implementing this National Training Plan (NTP). We are certain that by the end of 2020, teachers' professional and academic level will improve in China.

4.2 Overview of Primary Science Education Policy and Documents

In order to address the demands of scientific talent in contemporary societies, China has devised and implemented a series of laws and regulations regarding scientific education to ensure the effective development of the science curriculum. With the "Compendium of Curriculum Reform of Fundamental Education (1999)" implemented, the new curriculum reforms begun all over the country. Guided by this program, the Ministry issued the "Science Course Standards of Full-time state education (draft)" for Grades 3–6 as well as for Grades 7–9 in July 2001 (Ministry of Education 2001). The official media called this the first curriculum standards issued since the Communist takeover in 1949, and it held great significance for improving science education. The standards are set to cultivate scientific literacy as the primary purpose of the science curriculum. This will help guide teachers and students to question the traditional curriculum that has paid too much attention to knowledge and skills during teaching. Instead, the focus is now on the affective aspects, attitudes, and values so that students regard themselves as investigators of science during instruction. In addition, learners will make connections between what they learn in science lessons with the real world and experience or model some of the scientific habits/values of science (Poisson 2000). In 2002, the "Popularization of Science and Technology" law was issued indicating that China wanted to strengthen the popularization of science and technology to improve the scientific and cultural attributes of her citizens steadily. It means that scientific popularization has reached a new stage and a good atmosphere of respecting science and advocating for science education in the whole of society is formed.

In 2006, the document "Outline of the National Program for Long- and Medium-Term Scientific and Technological Development" was promoted officially to build an innovation-oriented country (Chinese State Council 2006). It means that construction of public scientific literacy had entered a new period of history, driven by the government, participated by all people. In the meantime, another paper printed and distributed by State Council "Outline of National Action Scheme of Scientific Literacy for All Chinese Citizens (2006–2010–2020)" hoped that by the end of 2020, public scientific literacy would have reached the level major developed countries reached in the early twenty-first century (State Council of the People's Republic of China, 2010). By the middle of the twenty-first century, each citizen in China would have basic scientific knowledge with scientific literacy improved significantly. Although this paper is about the improvement of scientific literacy, it has lasting implications for science education development in China, including implementation of young peoples' scientific literacy, STEM programs, and popularization of science equipment infrastructure projects. It has also stipulated the main tasks and mission of science education within elementary education, the training of science teachers, the importance of science education research, and the science curriculum and measures for content implementation.

If science curriculum standards have provided a detailed criterion for science curriculum, the report "Outline of National Action Scheme of Scientific Literacy for All Chinese Citizens (2006–2010–2020)" has then established the importance of science education in boosting science quality and education careers from a larger perspective. In 2010, an important report "Outline of China's National Plan for Medium and Long-term Education Reform and Development (2010–2020)" enacted by the State Council of China indicated that education in China was at a new historical starting point and about to enter a new stage for further development. This document advocated humanist education ideas for the first time and established a blueprint for the development of education in the next decade in China that greatly strengthen comprehensive education development. But it is very obvious that implementing changes from this education document in the coming 10 years does not clearly specify the role of science education in education in general or in broader society. Additionally, this does not relate to the two aims of the science curriculum or the report "Outline of National Action Scheme of Scientific Literacy for All Chinese Citizens (2006–2010–2020)," which is a big pity and deficiency.

4.3 Overview of Primary Science Education and the Curriculum

In 2001 the "Primary Science Curriculum Standards" was issued. Although new junior middle school science curriculum standards were revised and implemented in 2011, the revised primary science curriculum standards have been delayed. Hence, the science curriculum in China has followed the 2001 version, and its basic teaching ideas and aims are listed as follows:

- *Science education for all.* It means that every student possesses equal rights and chances to study science in the curriculum. The curriculum, teaching materials, teaching, and assessment are modified taking into account student differences.
- *Priority given to students' interests in science learning.* This clearly defines the role of the teacher in organizing students' science learning activities, namely, teachers play the role of facilitator and director in the science classroom, as well as partners in learning. During science learning, students must be given full rein to develop initiative, to solve problems by themselves.
- *Science inquiry for learning activities.* Science inquiry is not the only goal of learning science but also the content matter of science. The learning style advocated in science learning requires that students must actively engage in practical activities. Modeling how scientists investigate is one major goal of student learning in science.

- *Science teaching for both the needs of society and student development.* Science content for teaching should be close to children's daily life as much as possible and prepare for future developments in society.
- *Flexibility as special characteristic of the science curriculum.* It is reflected in many aspects of the science curriculum, such as the learning of content, teaching organization, assessment, etc. Students experience the science curriculum in the fullest sense when schools, family, society, and a variety of other resources help to richly enrich learning experiences.
- *Science teaching for scientific literacy.* The formation and development of scientific literacy is the biggest concern in the evaluation of the science curriculum. More attention should be paid to learning process, assessment methods, and more diverse indicators of literacy.

In terms of preparing primary science textbooks and other instructional media, it is based on the curriculum standards mentioned above. In consideration of the development needs of students from different regions, there are dozens of primary science textbook versions in China which reflect different styles. Currently, there are three publishing companies: Educational Science Publishing House, Jiangsu Education Publishing House, and Hunan Education Publishing House. The version published by Educational Science Publishing House is one of the most widely used teaching materials in 24 provinces and cities. In order for primary school teachers to make better use of teaching resources, these materials have supporting materials such as science teacher guidebooks, CD-ROMs on science teaching, and science teacher development and many others. The complete package has a total of 37 units (140 activities) in which 10 units are in the theme of life sciences, 11 in the physical sciences, 8 in earth-space sciences, and 8 in interdisciplinary units. Other than those mentioned publishing houses, China has other presses such as Elephant Publishers, Qingdao, and Guangdong Publishing house. Overall, there are a variety of teaching resources with different writing emphases.

Primary science begins from Grade 3: there are three lessons a week, and each lesson lasts for 40 minutes with one school-based science practical lesson to cultivate students' interest, inquiry abilities, and scientific literacy in science (Liu et al. 2012). In the forthcoming science curriculum standards, experiencing twenty-first-century competencies and core abilities is uppermost goals. Primary science classes may begin from Grade 1, as well as the corresponding revision of textbooks.

The number of periods of science in compulsory education has been fixed for Grade 3–6; class hours should be 7–9% of the total. It is intended that there should be approximately 30 hours of instruction per week, including 2–3 class hours of science. So from Grade 3 to Grade 6, 2–3 lessons for primary science are compulsory per week. According to a survey of western and northern rural schools in Guangdong Province (Zhang 2015), science classes in most schools are insufficient: 46.7% of schools have an hour per week, and 7.6% do not have any science. Most science lessons were also held in the afternoons where students' ability to concentrate is not high. If Guangdong as one of the more economically developed provinces in China faces such issues in science education for rural children, we wonder about other schools in the more rural regions.

Rural primary science education raises a number of worrying concerns such as:

- *Lesson time is not guaranteed.* Usually, most rural schools have a science period a week, but even this is not even guaranteed as it is often replaced by Chinese or mathematics teaching or just cut down to 10 min.
- *Science class without experiments.* The expository method is widely used in science classrooms without laboratory work.
- *A lack of committed primary science teachers.* Teachers are unwilling to teach, students are unwilling to learn, and parents refuse to cooperate. Science teachers in rural schools in China face many constraints and difficulties as well as lack of preparation. Most of them are unwilling to teach science, which is the main obstacle in the successful implementation of the primary science curriculum. Students generally express great interest in science and scientific knowledge, but due to the influence of classroom scripts, they slowly begin to lose a desire to learn science. Parents are influenced an exam-oriented system and fret constantly over grades in languages and mathematics and prefer children to learn practical skills. Parents in rural areas think science is like a game and are not willing to support buying some necessary materials for science activities.

Although rural primary science is not satisfactory, urban primary science has made much progress. Students can do inquiry activities in class, get knowledge about natural science, and do some investigations with the help of their teachers. So assessment of students' work is put on the agenda. Science literacy and science inquiry abilities of students are assessed in elementary science education. Assessment should guide the science teacher toward developing in a strong way hands-on and minds-on activities and not just limited to learning fragments of knowledge. It should assist students master science knowledge and enhance science literacy in observation, research, and doing experiments. How can we assess such difficult skills?

"Science inquiry ability" assessed through performance tasks. The test of scientific inquiry ability mainly aims to detect students' ability in hypothesizing, observing, experimenting, measuring, and giving scientific explanations and appropriate communication, which are at the heart of scientific literacy. Of course, in the initial stages of evaluation, science inquiry ability assessments can also encourage teachers to teach science through activities and make this way of teaching a norm. The selection of what to assess about scientific inquiry can be extracted from science activities in science textbooks, or teachers design it by themselves based on the primary science curriculum standards and requirement from several teacher guidebooks. To evaluate science inquiry ability, teachers can design a few science inquiry activities and allow students to attempt these independently. Through on-site observations, behavior records, explanations of the process, and scientific principles, teachers can evaluate students' practical skills, scientific reasoning, and creativity in making products. Alternatively, teachers can put forward a science inquiry task and assess the process from carrying out the hypothesis to completion, including knowledge of science inquiry and the science processes, methods of inquiry, and communication. Usually, teachers should build a detailed marking scheme for each practical task and illustrate a series of expected result, guidelines of general description of what to do, and information form before carrying out the on-site evaluation.

Listed below is an example of how to assess the inquiry process of making a simple electromagnet and determining its poles from Chinese science classwork.

Activity – make a simple electromagnet from the materials listed below, and determine its poles:

1. Materials: wires, iron nails, batteries, pins, and magnetic needles
2. Experiments and tests

    (1) Make a simple electromagnet and ensure that it works.

    (2) Use the electromagnet to attract the pins, and record the quantity of pins attracted in the table.

The number of pins attracted			
First try	Second try	Third try	Average

(3)Conclusion:____

(4) The purpose of repeating an experiment three times is____________

①Doing more for fun
②Reducing the errors in the data
③Making a comparison between data
④Getting more accurate data

(5) Determine the poles in the simple electromagnet below, and write down your answers.

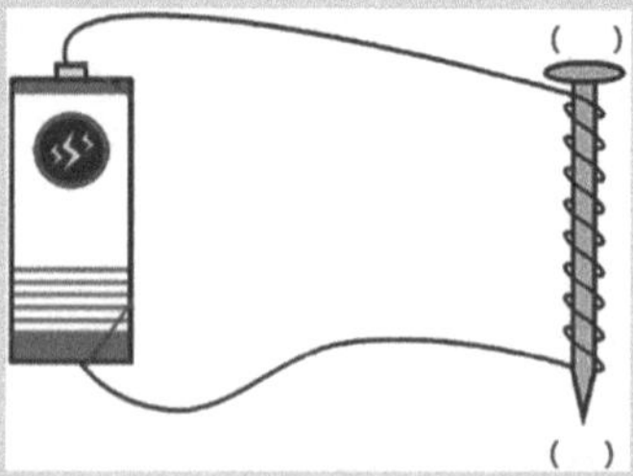

(6)Explain how you made this conclusion,

"Scientific literacy" assessed through written tests. Written assessments mainly tests students' understanding of science concept and mastery of knowledge. Assessment of this sort is through class tests or based on the science curriculum standards including themes in the life sciences, physical sciences, earth and planetary sciences, and the social and human sciences. It also includes conjectures and hypothesis at the heart of scientific inquiry, as well as planning, collecting data, reasoning, forming conclusions, etc. It is worth pointing out that evaluation of concepts is not just at the level of remembering and reproduction. On the contrary, applying what we have learned to analyze and solve problems, analyzing and processing a variety of information to access facts and evidence, and explaining the experimental result are emphasized. Also, applying knowledge is emphasized because many questions arise from students' everyday life. We can therefore examine their ability to extract, analyze, and process information in authentic contexts. This follows the curriculum idea "science comes from everyday life, teaching goes back to life."

Listed below is an example of a school-based assessment of the science concept of temperature:

Sheep Xi and sheep Mei want to measure the temperature of their house, so they buy two thermometers and hung it in their bedroom.

(1) This is Sheep XI's records.

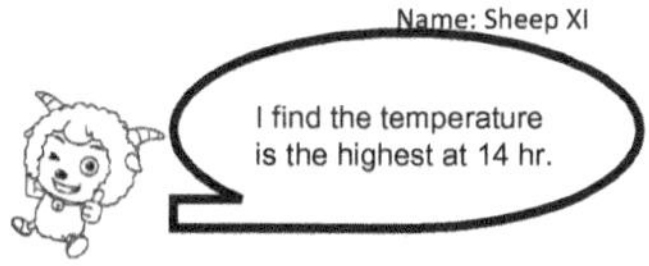

Name: Sheep XI date: January 1, 2016

time	6	8	10	12	14	16
temperature (℃)	3	3	8	13	16	15

(2) This is Sheep Mei's records.

Name: Sheep XI date: August 1, 2016

time	6	8	10	12	14	16	18
temperature (℃)	8	10	25	30	34	36	27

(3) Draw the line graph of temperature change in a day from the data of Sheep Xi and Sheep Mei.

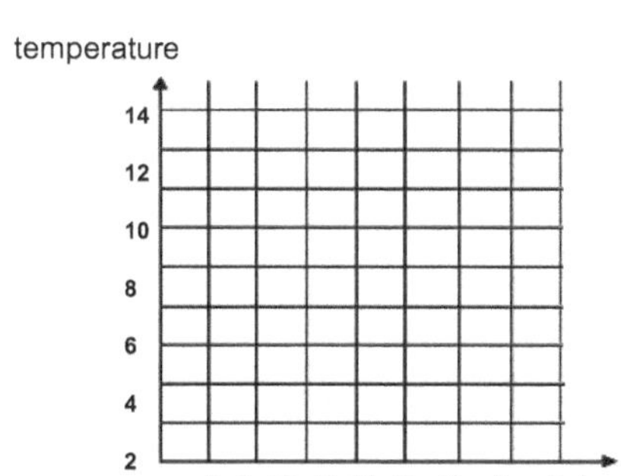

(4) Observe what you draw in 3, what else affects the temperature changes you see ?

Example of a Grade 5 science practical work in local schools

Observe changes of shadow in the sunlight

1. Find an object in school and mark its shadow in the morning. Mark it again during class break.

2. Fasten a pencil vertically in the middle of a piece of paper and mark the north-south direction. During the class break, draw the shadow of the pencil, and

Simulating the change in the position of the sun in class, and observe the changes of the shadow. (You can use a bamboo rod placed in a semi-circle above the stick and use a torch to mimic the sun. The sun moves in a semi-circle around the sky).

- What is the relationship between shadow direction and position of the sun?
- What is the relationship between shadow length and height of the sun?
- Predict the relationship between temperature change and change in sun position.

Write down and discuss the result of the changes in the data on shadows in your notebook.

4.4 Effect and Trends in Science Curriculum Reform in China

4.4.1 Policy Ideas for Guiding Student Development

Science curriculum policy changes in the future should emphasize the following three aspects as listed below:

1. Change the emphasis on abstract instruction and transmission of knowledge. In its place, stress in the curriculum the development of every students' physical and mental health, good moral character, and the desire and ability of lifelong learning.
2. Change the emphasis on disciplinary independence and lack of synergy. Conversely, strengthen integration, balance, and careful selection of content from the course itself.
3. Change the emphasis on the rigidity of the disciplines and the learning of transmitted content. Conversely, enhance the relationships between curriculum content, scientific and technological progress, and students' daily lives.

In short, based on the actual needs of students, we should pay attention to students' learning attitudes and emotional experiences (Wang et al. 2012).The subject status of students should be recognized and respected so as to respect the dignity of students.We should create a harmonious environment in teaching and encourage exploration and inqiry, developing students' innovation and practical abilities,making them konw how to learn.

4.4.2 Equality of Decision-Making in the Science Curriculum

With the enhancement of curriculum awareness, more and more people, organizations, groups, and other nongovernmental organizations are participating in curriculum decision-making with enthusiasm. Thus, this makes science curricular content reflects the interests of a broad mass of people. In the process of curriculum decision-making, besides the role of executive branches, government officials, and curriculum experts, some research institutions, institutions of higher learning, publishing division, social groups, and the media should either directly or indirectly be involved in science curriculum decision-making.

In addition, the status of teachers, parents, and students in science curriculum decision-making has been improved to a certain extent. This illustrates that the ability to participate in standards-based education and curriculum policy-making has broadened. Two broad types of groups or organizations exist in China; official groups have the legal authority to develop curriculum policies. Of course, some of them may be in fact subject to unofficial political groups or pressure groups including heads of government, legislators, administrators, etc. Unofficial groups refer to interest groups, political parties, citizens as individuals, etc., and their broad participation in science curriculum decision-making is a significant move toward equality.

4.4.3 The Function of Intensive Course Evaluation to Improve the Levels of Teaching

Teaching evaluation is an indispensable part of teaching activities, and evaluation can make teaching activities more scientific. Correct use of teaching evaluation is of vital importance to improve education quality. Primary science teaching is based on the inquiry model which is a widely promoted educational method in science all over the world. Teaching based on inquiry is not a simplistic teaching approach, instead inquiry is a major feature of science teaching that can be found through a variety of ways, such as teaching in experiment activities, survey, and scaffold instruction for argumentation (Yang 2013).

Inquiry-based teaching and traditional science education share some similarities, but in many ways they are different. Inquiry-based teaching is not just about processing or going through teaching resources, the main factor is for the students to participate in the identification of evidence, making judgments, reasoning logically, and thinking about how to interpret the data. successful inquiry-based teaching make students understand the natural world, develop inquiry skills and attitudes, have multiple perspectives toward solving scientific problems.

How we evaluate the degree of inquiry by teachers and students in this kind of environment requires a set of inquiry evaluation models. Evaluation of inquiry teaching falls into two kinds, assessment and test, and formative or summative methods can be used on both of them. Assessment is paying attention to teachers' teaching, and test refers to checking students' learning. Evaluators need to understand the behavior of teachers and the teaching process and to assess their quality. However, test of student learning is more multidimensional and comprehensive, and all aspects of their scientific competence have to be taken to make a comprehensive judgment. These should include information on the scientific concepts and linkages between concepts set forth in the standards, language and communication skills development, inquiry skills development, and scientific attitude development. Student learning test methods include formative and summative evaluation; formative evaluation should be dominant, and we have to consider a variety of ways and multiple evaluation methods.

4.4.4 Advocating Informal Primary Science Education

Regarding science activities outside school, science education is to be found in science exhibitions, science competitions, science and technology innovation competitions, construction of robots, various model design activities (such as aeromodelling, architectural models), summer camps, tech caravans, and so on. Some workshops also attempt to organize some classroom activities in relation to the primary science curriculum, such as doing interesting experiments, fun bingo games, etc. Howerve, we have to realise that informal primary education in China still stays at an exploration stage, and when we organize sepcial science activities, we have to be conscious of considering the goal of science education and students' needs (Guo 2014).

According to statistics from the China National Children's Center (CNCC) in 2012, 75% of students taking part in outside school informal education are primary children, and 19% are high school students. This shows that primary students are the key target groups involved in informal science education. Further, Fig. 4.4 details the ages of students involved in outside school informal science education; children aged 9, 10, and 11 are the most common, and as they grow older, fewer students participate in it.

Scholars in China generally believe that our science education tends to center on disciplinary themes and scientific fact that belong to lower cognitive levels. So advocating extending knowledge of science through these activities is getting more and more consensus. One way to go beyond low cognitive levels is the establishment of informal science education systems to focus on critical thinking and deep understanding of science. The other approach is to increase the history and philosophy of science (HPS) content in informal science education systems (Wan and Zhang 2013). Now some teaching materials flag out HPS content in some columns designed by editors in China, but HPS has yet to permeate informal primary science education. As mentioned above, primary students are the major target group in informal science activities, so more efforts should be taken to improve HPS education in primary students in practical ways. Some approaches to this are help-

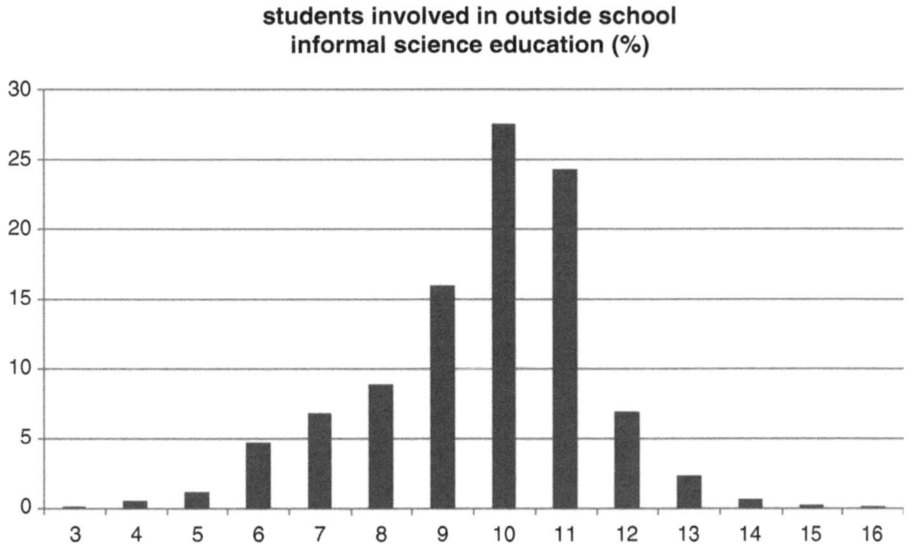

Fig. 4.4 Students involved in outside school informal science education (Data source: National Children's Center, China, 2012)

ing understanding science from HPS perspective, understanding the scientific spirit, and aspects of scientific work and ethics of science through the unfolding history of scientific development such as the many vivid stories of scientists' work. A thorough recognition of HPS in outside school science education will play a vital role in promoting science literacy.

4.5 Some Local Programs in Primary Science Education

4.5.1 "Learning by Doing (LBD)" Primary Science Reform Program

One well-known experimental project in China aimed at the improvement of primary science reform is the Learning by Doing (LBD) ("做中学") program. LBD was introduced from France and promoted by China Association for Science and Technology (CAST) and Ministry of Education that aimed to making students learn, inquire, and do science through hands-on activities and experiments at preschool and primary levels (http://www.handsbrain.com/). It was to allow preschool and primary students opportunities to personally experience, explore, and investigate in science as a teaching method. This is a national program of learning science that had been proposed as part of international reform efforts in science education. The LBD has a clear educational philosophy, objectives, and guidelines with a realistic and contemporary mode of implementation. The LBD is an educational philosophy but also a method or process of education. Through LBD, it can improve children's scientific development taking into consideration their personal characteristics but also embodies the purpose and spirit of quality science education. Result of a 3-year research on LBD indicates that the mode can not only develop children's sense of science and ability but also cultivate children's habit of team work (Lai and Ding 2005). Features in LBD include:

- *Respect of differences among every child.* Every child in China should receive an equally good science education. In the LBD program, we encourage every child to work based on their interests, desire, and abilities to explore and learn science.
- *Lay a foundation for lifelong learning and learning to live.* Children are born with curiosity and enthusiasm to explore as much as scientists, we have to give serious attention and vigorously support them to maintain and develop this. LBD encourage childen to do hands-on activities, express their views, have a sense of responsibility, try to solve real-life problems. The LBD form of education will improve the scientific literacy of children, for their future to adapt to society and lay the foundation for a good life.
- *Environment and life are where teaching resources come from.* We want to focus on and pay attention to children's interests in life, to address their needs and use them as an important source of teaching scientific content. We will select accessible and educational value of things from the lives of children, as their starting operating materials for practical inquiry. Based on the primary science education goals formulated by the state, we will be flexible in designing, to highlight customization and reflect local characteristics of the case. In this form of scientific education, what the children learn are not random, fragmented knowledge, but they need to master the necessary links between scientific concepts and to develop initial scientific and cultural awareness.
- *Guide children to take the initiative to explore and experience the discovery process.* Core principle of the LBD program is to make children experience process of scientific inquiry, such as problem finding, predicaiton, recording information, and communication with each other. Through this process, children's preliminary inquiry ability and problem-solving ability are expected to grow.

The LBD science program is a teaching reform activity in view of the reality of disadvantages present in education in China. Also, it is a learning method that fully considers children's physical and mental development characteristics. Based on hands-on activities, the LBD stimulates students' initiative, pays attention to the scientific method of learning, and carries out of inquiry under the guidance of theory. Some approaches to implement it smoothly and rapidly include design teaching materials, research in teaching practice, teacher training workshops, and international cooperation supported by network.

The significance of LBD lies not only in improving scientific understanding and exploration for future middle and high schools; it can deeply affect the behavior and ways of thinking of the child. It can contribute to loving science, doing science, and immersing in the culture of science. We feel it is bound to improve and perfect the education system, to develop and enrich science education in our schools and society, and to have a profound impact on the quality of our nation.

But some factors also restrict its development, such as low level of science literacy of teachers. According to some investigations, the work burden also makes teachers carry out few experimental work. Human resource, material resources, and network resource cannot catch up with the rapid needs of children; lots of teaching materials have been used for many years and obviously outdated. LBD science program has implemented for 15 years; although some problems have surfaced, we believe that it must have far-reaching effects on primary science education in China.

4.5.2 "Learn to Think (LTT)" Intervention Program

Based on Piaget's cognitive development theory, Vygotsky's theory of social constructivism, and Lin's theory of learning and development, the "Learn to Think" (LTT) activity curriculum has been designed with a series of learning situations that are planned to be conducive to the development of students' thinking methods or strategies and improving the quality of thinking (Hu et al. 2011).

Cognitive research takes account of affective and social aspects, and interest and motivation play an important part in thinking because thinking requires effort. Lin proposed that non-cognitive factors, including motivation, interests, etc., have a direct effect on cognitive processes. Therefore, the LTT thinking curriculum must aim not only just to teach skills and knowledge but also to develop interest and motivation in their use. All aspects of activity selection, from choosing activity content, materials, and situations to producing cognitive conflict, teacher–children social construction, and reflection or transfer, are focused on stimulating children's learning interest and motivation, encouraging children to explore learning methods and strategies, and staying positive and active in the acquisition of scientific thinking.

LTT has five characteristics (http://www.learntothink.cn/):

1. Suitability. The difficulty of each activity is in the "zone of proximal development," and it is not only suitable for but also can promote the development of thinking ability of students.
2. Systematization. The whole project covers almost all the thinking methods appropriate for students across the eighth grades from the 1st grade of primary school to the 2nd year of secondary school. Each thinking method appears in different grades, and the thinking difficulty increases with grade. The whole curriculum trains the thinking ability of students step by step in a spiral manner. For example, categorization is arranged as follows: Grade 1 (age 6 years), categorize shapes and objects/things; Grade 2, categorize unfamiliar and living things and cross categorization; Grade 3, use categorization in problem-solving tasks; Grade 4, categorize based on purpose and the limitedness of categorization; and Grade 5 (10 + years), categorize three-dimensional graphs and recognize the multidimensionality of standards.

3. Diversity. Activities of LTT include thinking method training (such as "analogy, compare, analysis and generalization), space cogniton, problem finding, problem-solving, scientific inquiry story inventing, and so on.
4. Elicitation. Each activity is planned to induce some cognitive conflict in students, encouraging them to think actively.
5. Gradualness. The arrangement of activities accords with the character of student thinking, from simple to complex, from shallow to deep, from easy to difficult, step by step, to enable students improve the ways of thinking gradually.

According to teaching practice, LTT has got lots of research achievement under efforts of participated teachers and students. Researches find that LTT has deep influence to primary students' thinking ability. Innovation quality is also improved in LTT curriculum; it can enhance students' creative tendency and abstract thinking ability effectively.

4.5.3 STEM+Primary Science Research Project in Shanghai

As a pilot education reform project in Shanghai, the STEM+ primary science research project developed by Stanley science education research center has been implemented in some primary schools(http://www.bestway-rtc.com/). It started in October 2014 with a celebration of some teachers getting the qualification of starting the STEM+ curriculum. The symbol "+" means that this reform curriculum not only contains science, technology, engineering, and mathematics but also pays attention to humanities and arts; it is the dynamic integration of the two parts.

The STEM+ project aims at cultivating students interdisciplinary ability and integrated big ideas, so the symbol "+" can be regard as adding science content, but more importantly, enhancing education concepts is the final goal.

Some of its theoretical ideas are listed below:

- TEM+ is a curriculum based on project learning. In STEM+ activities, doing some projects is not the main purpose, and understanding concepts and enhancing scientific abilities are the fundamental goals; the curriculum project is the "carrier" toward finishing the two goals.
- STEM+ is a kind of new science teaching mode. Classroom teaching has changed in TEM+, teacher-centered teaching mode gets challenging, and teachers have to make lots of time to guide the learning process. For example, in a project of making a rain vessel, two-thirds of time accounts for brainstorm, and the teachers' mission is to help students to make a perfect vessel measuring the rainwater.
- STEM+ helps to cultivate ability. The goal of STEM education is to nurture STEM talent, so this reform project aims to make more students interested in STEM disciplines in the future. In STEM+, curriculum content is always associated with hands-on ability and thinking training; when the two parts are considered sufficiently, it means a new level of moral education is also reached.

This project is underway, but some questions have come to light, such as lack of teachers, too much relying on executive decision and lack of self-development, impractical teaching materials, etc. But what we also can see is more doing of hands-on works in the classroom; this is the baseline of doing STEM in primary. So we hope this project can effect true change in the traditional teacher-centered modes of teaching in China and reflect STEM ideas, further integrating STEM and art, culture.

4.6 The Method of Primary Science Teacher Education

4.6.1 Training Programs of Preservice Primary Science Teachers

Since entering the new millennium, preservice education of primary school teachers begins to change from being in a vocation to being a professional. So far, more than 60 institutions of higher learning by the Ministry of Education have approved the establishment of a Science Education department. In the past 10 years, four modes have emerged in training preservice primary science teachers (Ding 2011):

Mode One: Undergraduates from Elementary Education Universities Here a Normal College has set up a separate primary education program. Or, former Secondary Normal Colleges or other teacher education institutions have been upgraded to teach at the undergraduate level. Primary school science teachers learn similar courses in language, mathematics, and English, but undergraduate institutions are different. Its special characteristics are training full-time primary school science teachers with higher education scientific literacy and science literacy in science class professional curriculum to better reflect the structure of knowledge and ability structure of full-time science teacher.

Mode Two: Undergraduates from Natural Science Departments of Education Universities This mode is typically for those specializing in primary education in Shanghai Normal University. Nanjing Xiaozhuang University also adopts this mode to cultivate pre-service elementry sciece teachers. It divides primary education into the "liberal arts" and "big science" categories. Primary science teacher training starts from "big science" (mathematics and science) direction. Compared with the first model, training objectives here focuses on both disciplines of math and science, but mathematics curriculum courses are more numerous than those in the science curriculum. Educational courses are present (see Table 4.1), but not as many mathematics and science courses.

Table 4.1 Course classification of primary science teachers training in NTP (Data source: "National Training Plan (NTP)" curriculum standard, Ministry of Education, China 2012)

	Programs		
Course contents	Primary and secondary teachers training showcase projects	Key teachers' training projects in mid-west rural	National nursery teachers' training program
Professional concept	10%	10%	5%
Professional knowledge	40%	35%	40%
Professional competence	50%	55%	55%

Mode Three: Science Education Undergraduates from Science Departments of Universities Since 2002 the Ministry of Education has been promoting the new basic education curriculum in order to facilitate a junior high school integrated science curriculum and set up a new profession. In China, high school doesn't set up integrated science curriculum instead with subject curriculum such as biology, chemistry, geology, and physics except province Zhejiang and Shanghai. So Science Department in Normal Universtiies are intended to cultivate high school science teachers, such as the College of Chemistry and Life Science in Zhejiang Normal University trained Bachelor of science graduates. Its goal is to train qualified integrated science teachers and enhance the teaching quality of high school integrated science.

Mode Four: Colleges for Teacher Training In this pre-service teacher training mode undergraduate students study 3 years basic knowledge of teaching, then 2 more years is add to learning college science knowledge and teaching knowledge. It shapes solid teaching basic skills and equipped for primary math, Chinese, and science teachers. **The model schools are upgraded from the original Secondary Normal College, which generally focus on developing primary language (Chinese), English, and mathematics teachers and other disciplines, but there are some schools to establish a culture of Natural Sciences full-time primary school science teacher.

Different patterns of elementary school science teacher training mode reflect different routes of curriculum reform in elementary education. Also, educators' view to primary science discipline and primary science teacher changed over time. More and more educators realize the importance of cultivating professional primary science work teams.

4.6.2 Professional Development of Primary Science Teachers

Primary science teachers' continuing education gradually came into public view with the "National Training Plan (NTP)" implemented by the Ministry of Education and Finance at the beginning of 2010. NTP is the implementation of the document "Outline of China's National Plan for Medium and Long-term Education Reform and Development (2010–2020)" to strengthen the professional development of teachers in kindergartens and primary and secondary schools. The emphasis of NTP is rural techers and encourages all teachers to improve their teaching ability and professional level. Three projects which might be an effective means to improve teaching quality are in implementation to promote the popularization and balanced development of compulsory education (Ministry of Education 2012). These are the "Primary and secondary teachers training showcase projects" which are directly supported by the Ministry of Finance to train key teachers of primary and secondary nationwide, "Key teachers training projects in mid-west rural," and "National nursery teachers training program" which are supported by central special transfer funds for training teachers in the middle or western parts of China.

As far as the training program of primary science teachers is concerned, training courses include professional concepts (ethics, professionalism), expert knowledge of teaching (subject knowledge, pedagogical content knowledge, general knowledge), and professional competence (teaching design, teaching implementation, evaluation, research, use of educational technology, development and utilization of curriculum resources, classroom management, etc.). Their main content areas (in %) are listed in Table 4.1.

4.6.3 Curriculum Implementation

Curriculum implementation is mainly divided into two stages, namely, centralized training and decentralized training. In centralized training, teachers involved must finish some professional knowledge courses such as intensive training in the science curriculum, case analysis, and discussions about teaching idea. In decentralized training, the purpose is start to deepen and practice what was learned in centralized training courses and to carry forward more training to adapt primary science teaching for individual differences.

1. Centralized Training

In accordance with the needs of primary science teachers who generally lack theoretical content knowledge, combing lectures and seminars are good ways to deepen their understanding of the professional content through thinking, discussions, and listening. But lectures and seminars are not the main approaches for we advocate participatory-oriented, case-oriented, and workshop-oriented ways to improve training efficiency too.

Participatory-Oriented Participatory training involves the full participation in the context; teachers experience participating in the training initiative, cooperate, and explore learning process, so as to achieve learning the concepts firsthand. There will be changes in attitudes and behavior, and they will be able to learn knowledge and methods for their own work training activities. Scientific inquiry is supposed to be one of the main teaching methods that elementary school science teachers should use, but how to effectively implement scientific inquiry generally causes confusion. Through training primary school science teachers can experience the process of inquiry teaching. From these three forms of active participation in an atmosphere of scientific inquiry, teachers' teaching ability and levels will gradually improve.

Case-Oriented Only when teacher's in-service training is being closely combined with their classroom activities can it improve teaching ability authentically. So case-oriented training focuses on analysis of primary science teaching case, ways of applying teaching strategies, gradually increasing teachers' teaching competence in teaching classroom observation, seminars, their reflection, and practice of teaching. Also, case-oriented training generally is directed toward teaching problems in science classroom, reflecting more relevance and practical operability.

Workshop-Oriented Workshop-oriented training is based on real teaching scenarios, and the participants involved in the process of dialog can communicate with each other, share ideas, investigate and analyze, make proposals or plans, and discuss how to promote the program, how to put into practice, and how to reflect, improve, and perfect the program. It is generally based on real teaching situations, around specific themes. The participants interact with excellent subject teachers and subject specialists and develop action plans and improve teachers' practical ability in practice and reflection.

2. Decentralized Training

In this second period of training, science teachers will continue to deepen the training programs through the concrete practice of science teaching. Two ways are adopted: school-based training and Web-based training.

School-Based Training School-based training begins in primary school, based on teachers' teaching practice to improve their teaching ability. As short-term training hardly improves teaching competence effectively, follow-up school-based training makes training in real circumstance become a reality. In daily teaching practice, teachers put to use what have learned in previous intense training and then communicate with each other to get more consistent opinions about teaching design, teaching philosophy, and teaching strategies.

Web-Based Training The Ministry of Education issued the "network alliance plan about accelerating teachers' education" document in 2004 to enhance teacher training in teachers' network alliance. It announced that teachers' education network in each city should be used fully to construct rich resource for science teachers' continuing education, eventually providing strong support for primary science teachers' professional development. Some subprojects have come into service such as experts replying online, asynchronous peer interaction, video lectures, online learning, online assignment, and so on. Web-based training provides a good platform for teachers and experts to share resources, cooperate and communicate, give opinions, and so forth.

4.7 Science Curriculum Management System and Students' Feedback to It

Three levels of science curriculum management systems have been built in China, namely, the national, local, and school science curriculum. These three levels align curriculum to meet local, school-based and individual situations.

The national science curriculum is regulated by the national education administration. It reflects the state's intention to define what students should obtain after receiving compulsory education. Development of the national curriculum is based on educational objectives in different stages to set the curriculum standards and writing of textbooks. The national curriculum is the main instrument of national science compulsory education curriculum frameworks, which play a key role in school science curriculum implementation. So, national science curriculum serves as important factor in determining national science education quality.

The local science curriculum is planned by provincial departments of education or local education authorities/departments to develop curriculum that reflects indigenous culture or science and technology development. It takes full advantage of unique native/local education resources, reflects regional characteristics of basic science education, and is adaptive to contexts. With the development of local science curricula, we begin to see more regional science projects that cultivate students' feelings of their sense of place.

School science curricula are mainly compiled by school science teachers. After evaluating students' needs from teaching experiences, it makes full use of local community and school science curriculum resources. Mindful of the school context, teachers plan diverse science courses for students to choose. School science curricula fully respect and meet students' personality, especially paying attention to the content that cannot be achieved in national science curriculum. School science curriculum can change over time flexibly and diversely to teachers' and students' need. It is an integral part of national science curriculum plan.

4.8 Distinctives and Changes in the New Elementary Science Learning Standards

In February 2017, the Ministry of Education in China officially released the new *Compulsory Education Primary Science Curriculum Standards 2017* to be implemented from the fall semester. The most obvious change in the new standards was to set the start for learning science at Grade 1, an earlier start of 2 years as compared to the previous curriculum. The new curriculum standards regard science learning in primary schools as a fundamental school subject (一门基础性课程). Thus, the Ministry has explicitly requested that all schools ensure that children in the lower grades can experience science teaching.

The new standards have some of the following key features:

Assimilate Ideas and Outcomes of Learning Progressions Learning progressions (进阶) describe pathways of (science) conceptual development that occur over a period of time in age-appropriate ways. It takes into account the student's age, his or her learning characteristics, and cognitive development as well as considers the special characteristics of the scientific discipline itself. The new curriculum standards will be divided into three sections in primary schools: Grades 1–2, 3–4, and 5–6. A spiral curriculum will be adopted whereby the same topic or concept will be addressed at different sections with appropriate levels of demand. In this way, science teachers are clearer regarding the cognitive demands of the same concept across different sections and be aware of the extent of what to teach. This point has enormous influence on the preparation of science textbooks as well as the assessment of scientific literacy among students (Table 4.2).

Reflect STEM Concepts Many countries advocate STEM (science, technology, engineering, and mathematics) as a modern understanding of science education. Theory in STEM education emphasizes practices, stressing the integration of different disciplines. For the first time, the new curriculum standards include "technology and engineering" content as well as materials, life sciences, earth, and space science

Table 4.2 A sample description of learning standards (scientific knowledge) from the *Compulsory Education Primary Science Curriculum Standards 2017*

	Goals of scientific knowledge across sections		
Domain	Grades 1–2	Grades 3–4	Grades 5–6
Materials	Observe and describe the basic characteristics of common objects; identify common materials in daily life; know common forces	Measure and describe the characteristics of objects and their properties of the material; describe the movement of objects; know the applications of forces (to understand the different forms of energy)	Understand changes in common substances; know the conversion between different types of energy
…	…	…	…

that together comprise four major domains of study. Engineering design forms the core of STEM education; therefore "technology and engineering" in the new curriculum standards essentially reflects the heart of STEM education.

The original curriculum standards also provided a number of STEM interdisciplinary content, mainly embodied in a number of engineering design activities to help students learn through doing creative tasks. The new standards will strengthen engineering design and inquiry activities, to focus on "hands-on" and "Learning by Doing" activities to develop the habit of problem-solving among students.

It can be said that this is a fundamental change in the design activities from the former science curriculum. Previously, design activities were few, the ability to learn engineering thinking was weak, and whatever attractiveness of the technical task was small. Now the engineering design and technical application activities under STEM concepts are more organically unified from the perspective of knowledge integration and the use of methods and tools in the creative design process.

Highlight Learning the Core Ideas Another feature of the new curriculum standards is the core ideas of science. These assist in the integration of subject knowledge, deep exploration of important principles, development of students' deep understanding, and promotion of students' participation in engineering practices and scientific literacy. In the case of primary school science education, it is more important to stimulate students' interest in science and promote meaningful learning rather than to pursue the integrity and systems of knowledge. The new curriculum standards in the four areas of knowledge contain 18 core ideas from science, technology, and social and environmental content spread throughout the three sections as shown below.

Domain	Key core ideas
Materials	Objects have specific characteristics; materials have specific properties
	Water is a common and important compound
	Air is a common and important mixture
	The movement of an object can be described by its position, speed, and direction
	Forces acting on objects can change their shape and movement
	Movement, sound, light, heat, electricity, and magnetism are different forms of energy
Life science	There are different kinds of creatures on earth
	Plants can adapt to the environment and create and obtain nutrients to maintain their own survival
	Animals can adapt to the environment and survive by acquiring nutrients from plants and other animals
	A body has a number of systems, and the division of labor with a system maintains life activities
	Both plants and animals reproduce offspring so that they can be passed down from generation to generation
	There is an interdependence between plants and animals and between animals and plants and the environment

Domain	Key core ideas
Earth and space science	In the solar system, the earth, the moon, and other planets are moving regularly
	There are soil, water, creatures, and rocks on earth. The crust, mantle, and core are inside the earth
	Earth is the home of human existence
Technology and engineering	In order to produce objects and make life more convenient, fast, and comfortable, humankind has created a rich and varied technological world
	Invention in the core of technology that involves the use and transformation of nature
	The key to engineering is design; engineering is the use of science and technology to design, solve practical problems, and manufacture products

4.9 Closing Thoughts

China's vast geographical size, coupled with various educational policies and funding issues, has perhaps led to the uneven development of primary science education in the country. With large differences in urban and rural primary science education, the latter is a government priority for improvement. However, there has been general progress following waves of education reforms in China: school science curriculum goals have linked with wider literacy goals in the nation to improve students' science literacy; science teaching is shifting to core abilities to cultivate subject matter core knowledge, and ways of assessment are getting more diverse. Some local programs such as LBD and LTT have gotten more influential in enhancing students' ability in science thinking and practice, but their sustainability can be further improved. Primary science teacher professional development becomes an overarching aim; some programs in NTP have helped integrated progression in teaching beliefs, PCK, practical knowledge, and inquiry teaching ability. Reforms in the future will also be policy oriented for guiding student development, decision-making in science curriculum will be more equitable, course evaluation will be more intensive or holistic to improve teaching, and informal primary science education will be advocated.

We appreciate that primary science in China has seen many comprehensive changes in curriculum goals, content, teaching, and evaluation through learning from international science curriculum reform experiences. We recognize that the primary school science curriculum system has been established to improve science literacy in the early part of this century. While there is no doubt that primary science in western countries has become a basic or required subject, in China primary science has not obtained such equal status in educational fields or Chinese society. This limits its development we feel and also affects the improvement of primary science overall. In the meantime, as policymakers and researchers struggle to understand the development of primary science education in China through basic research in learning psychology in the discipline, more questions will certainly be exposed concerning curriculum, content design, and teaching strategies. These need to be confronted head-on in the future, which will present many difficult challenges for us.

References

Adelainde. (2009). Curriculum reform and 'quality education' in China: An overview. *International Journal of Educational Development, 29*, 241–249.

Chinese Association of Science and Technology. (2002). *Report on Chinese public scientific literacy in 2001*. Beijing: Science Popularization Publication.

Chinese State Council. (2006). *The programme for national science literacy promotion action (2006–2020)*. Beijing: The State Council (in Chinese).

Ding, B.-P. (2011). Science teacher education for primary schools in China: Status quo, problems and recommendations. *Contemporary Teacher Education, 6*, 1–7 in Chinese.

Guo, G. Z. (2014). *The theory and practice study of informal science curriculum and its development* (Unpublished PhD. Thesis). Northeast Normal University, China. (in Chinese).

Hu, W.-P., Adey, P., Jia, X.-J., Liu, J., & Zhang, L. (2011). Effects of a 'learn to think' intervention programme on primary school students. *British Journal of Educational Psychology, 81*, 531–557.

Lai, F., Liu, C.-F., et al. (2014). The education of China's migrant children: The missing link in China's education system. *International Journal of Educational Development, 37*, 68–77.

Lai, X.-L., & Ding, Z.-Y. (2005). Learning by doing': A form of science education for children. *Educational Research, 6*, 89–93 in Chinese.

Liu, X.-F., Liang, L.-L., & Liu, E.-S. (2012). Science education research in China: Challenges and promises. *International Journal of Science Education, 34*, 1961–1970.

Ministry of Education. (2001). *The outlines for curriculum reform in elementary and secondary education (experimental version)*. Beijing: Ministry of Education of China (in Chinese).

Ministry of Education. (2012). *'National Training Plan (NTP)' curriculum standard (experimental version)*. Beijing: Ministry of Education of China (in Chinese).

OECD. (2013). *Lessons from PISA 2012 for the United States, strong performers and successful reformers in education*. From https://doi.org/10.1787/9789264207585-en.

Poisson, M. (2000). *Science education for contemporary society: problems, issues and dilemmas. Final report of the international workshop on the reform in the teaching of science and technology at primary and secondary level in Asia: Comparative references to Europe*. From http://www.ibe.unesco.org.

State Council of the People's Republic of China. (2010). *Outline of national action scheme of scientific literacy for all Chinese citizens (year 2006, 2010 and 2020)*. Beijing: State Council of the People's Republic of China (in Chinese).

Wan, D.-S., & Zhang, H.-X. (2013). Predicaments of the development of informal science education in China and its path choice—Based on the experience of the US. *Education Science, 8*, 6–11 in Chinese.

Wang, S., Liu, X.-F., & Zhao, Y.-D. (2012). Opportunities to learn in school and at home: How can they predict students' understanding of basic science concepts and principles? *International Journal of Science Education, 34*, 2061–2088.

Yang, Q. -L.(2013). Discussion on suitableness of reform for curriculum and instruction in basic education. Education Research Monthly, 11, 3–8 (in Chinese).

Yang, J., Huang, X., & Liu, X. (2014). An analysis of education inequality in China. *International Journal of Educational Development, 37*, 2–10.

Zhang Y. -P. (2015). Review on education in rural primary school science. Journal of Nanjing Xiaozhuang University, 3, 38–42 (in Chinese).

Chapter 5
Primary Science Education in Taiwan

Ying-Tien Wu, Li-Jen Wang, Teng-Yao Cheng, and Zong-Rong Yang

5.1 Introduction

National competitiveness and economic growth of Taiwan mainly rely on the development of its science and technology industries. Because of this, a highly skilled workforce in science and technology will have a great impact on Taiwan's economy. In the knowledge economy which is believed to characterize the twenty-first century, preparing citizen literacy in science and technology plays an even more important role than ever. In the middle and long term, and taking into account local society and culture, science education in Taiwan has been under continuous reforms over the last 15 years. This chapter thus aims to describe past changes as well as the current status of primary science education in Taiwan. We begin with an overview of science education in Taiwan followed by an overview of the local science curriculum and textbooks, the focus of current primary science teaching, primary science teacher education and professional development, and finally assessment issues in the subject.

Y.-T. Wu (✉) · L.-J. Wang · T.-Y. Cheng
Graduate institute of Network Learning Technology, National Central University, Taoyuan City, Taiwan
e-mail: ytwu@cl.ncu.edu.tw

Z.-R. Yang
Wong-Zih Elementary School, Taichung City, Taiwan

Y.-J. Lee, J. Tan (eds.), *Primary Science Education in East Asia*, Contemporary Trends and Issues in Science Education 47,
https://doi.org/10.1007/978-3-319-97167-4_5

5.2 Overview of Primary Education in Taiwan

According to the National Constitution, all citizens in Taiwan are required to receive 9 years of Compulsory Education where children receive their primary education beginning at the age of 6. In Taiwan, the establishment of primary schools is district-based, and there are no schools especially established for aboriginal or rural children. Most of the primary schools here are public schools and some private institutions in every city or county. Both types of schools are however centrally monitored by the Department of Education in their respective city or county government. For example, all primary schools (public or private) are controlled by the Department of Education in the Taipei City Government. The education ministry in Taiwan supervises all the departments of education in the government of every city or county.

In recent years, in order to drive educational innovations, more and more experimental primary schools have been established in Taiwan. Most of them were originally public primary schools. The establishment and management of these experimental schools could be “state-run” (i.e., an experimental school established and managed by the government), “privately managed public-established school” (i.e., an experimental school established by the government but managed by a private organization), and “private-run” (i.e., an experimental school established and managed by a private organization).

The typical primary school day runs roughly from 7:50 AM to 4:50 PM (seven lessons per day, 40 min per lesson). However, there are different ranges of school hours for different grades. For example, for 1st and 2nd graders, they have one full school day and four half days in a week; for the 3rd and 4th graders, they have two full school days and three half days; for the fifth and sixth graders, they have four full school days and one half day. As Wednesday afternoon is a compulsory period time of teacher professional development for primary teachers, there is no class for 1–6 graders during this period.

Regarding the implementation of science courses, there is no science course in the first two grades. However, basic science, as the content listed in national science curriculum standards, is taught in the “Life Course” (two lessons per week), which is an integrated course about life education. In 3rd–6th grades, science (termed “Nature and Technology Course”) is formally implemented and is taught for three lessons weekly.

Normally, science teachers heavily rely on textbooks and workbooks in their instructions. Hands-on activities are also carried out with basic laboratory equipment. Instead of purchasing laboratory equipment with the tight school budget, primary schools often acquire free basic laboratory equipment from textbook publishers. In science classrooms, students are not streamed according to their science ability. Most of the students attend cram schools after school. Traditionally, no science course is provided in cram schools, and students just do drill practices related to their school science courses in cram schools. However, in recent years, some cram schools have started to provide lessons of science experiments or science projects. Also, some cram schools specialize in offering science courses for primary learners.

5.3 Overview of Science Education in Taiwan

5.3.1 Science Education White Paper and Report on Scientific Literacy

To respond to the huge challenges of maintaining (inter)national participation in the knowledge economy, the first National Science Education Conference in Taiwan was held in 2002. This conference was organized by nongovernment organizations, government officials, and scholars. After the conference, the "Science Education White Paper" which condensed a common consensus from the participants was formulated and then disseminated. In this document, the vision of science education in Taiwan was articulated as "Science education is for all citizens." The goals of science education were to cultivate citizens' scientific literacy, to develop thier creativity and ability to innovate, and to raise their awareness of global issues. (Ministry of Education and National Research Council 2003a, b). Since then, the goals of science education in Taiwan have changed from mere training of professionals in science and technology careers to developing all citizens' scientific literacy. In other words, science education in Taiwan was transformed from an education for elites to education for all (Cheng 2005). Given these developments, we believe that science education in Taiwan is on the right track.

In recent years, with the globalization of the economy and rapid advances in information technology, Taiwan has faced fierce and multiple competition from developed countries and developing countries. However, Taiwanese students' performance in international assessments in science and mathematics does not seem to have an advantage compared with nearby Asian countries, such as Japan, Korea, Hong Kong, Singapore, and China.

To overcome such challenges, the Ministry of Education (MOE) in Taiwan established the Office of Promoting National Literacy. This office promotes the development of five important literacies for citizens, including languages (Chinese and English), mathematics, science, digital, and arts. In addition, this government office announced a new report on scientific literacy (http://literacytw.naer.edu.tw/data/cht/20140430/20140430sxaero.docx) that included three aspects relevant to science education:

1. An understanding that scientific inquiry is the means whereby human beings can explore knowledge and create civilized societies
2. An ability to understand scientific knowledge and to apply it to solve problems in daily life with technological and digital tools
3. Willingness to participate in discussion or argumentation on scientific issues, have reasonable doubt in explanations of scientific phenomena, develop the habits of reflection and critical thinking, and finally to become a global citizen who is responsible for the natural environment and public resources

5.4 Overview of Science Curriculum and Textbooks in Taiwan

5.4.1 From Curriculum Standards to Curriculum Guidelines

Before 2003, the implementation of elementary science education in Taiwan was based on the "National Primary and Secondary School Curriculum Standards." In this document, primary science education (beginning from the first grade and ending at the sixth grade) and secondary science education (from the seventh grade to the ninth grade) were planned separately. During the period, primary science education was implemented as a stand-alone subject, and the acquisition of core scientific knowledge was the major educational goal.

In 2003, a new curriculum guideline for primary science education in Taiwan was announced. The present curriculum guideline in primary science education (Grades 1–9 Curriculum Guidelines in Nature and Technology Learning Area) was announced in 2008. It was slightly reformulated based on the earlier guideline announced in 2003. Unlike primary science education and secondary science education that were planned separately in the "National Primary and Secondary School Curriculum Standards," science education was now coherently planned from the first grade to the ninth grade in the present curriculum guideline. According to the present curriculum guideline, nature, science, and technology are all integrated in the curriculum. The main educational goals of the Science and Technology Learning Area at Grades 1–9 are organized as follows (Chiu 2007, p. 309):

1. To cultivate an interest in and a passion for science inquiry and the habits of active learning
2. To acquire methods of inquiry and a basic competence in learning science and technology and to be able to apply one's learning to daily life
3. To cultivate a loving environment, to treasure resources, and to respect life
4. To cultivate competence in communication, teamwork, and getting along with others harmoniously
5. To cultivate independent thinking and problem-solving and stimulate their potential
6. To be aware of and explore the interactive relation between humans and technology

In the present curriculum guideline, the basic competences that citizens should acquire after instruction in the Science and Technology Learning Area are emphasized and include (Ministry of Education 2003b, p. 3):

1. Process skills: Mental processing ability in the processes of scientific inquiry.
2. Knowledge of science and technology: Cultivation of science concepts and ideas about technology.
3. Nature of science and technology: Science is verifiable and technology is practical and usable.

4. Development of technology: Able to understand how science is advanced and how technology is developed.
5. Science attitudes: Appreciate seeking truth from facts, and feel the beauty of science.
6. Habits of thought: A holistic ability to think critically and solve problems as well as the ability to integrate information.
7. Applications of science: The ability to apply scientific knowledge and the methods of inquiry.
8. Design and production: The ability to use personal or group creativity to produce technological artifacts.

In general, curriculum reforms are conducted in Taiwan every decade. Since 2014, the extension of the Compulsory Education from 9 to 12 years of school has been implemented in Taiwan. Accordingly, the Ministry of Education of Taiwan has started formulating a new curriculum guideline in 2014. It is estimated that new curriculum guideline will be implemented in 2019 and would emphasize the cultivation of scientific literacy. Science instruction would also be taught from Grades 3 to 12. Different from the Grades 1 to 9 Curriculum Guidelines in Nature and Technology Learning Area (i.e., Learning Subject), science will now become a stand-alone subject in the new curriculum guideline (http://www.naer.edu.tw/files/15-1000-7944,c1174-1.php?Lang = zh-tw).

According to the draft document, the educational objectives of science learning are listed below:

1. To inspire enthusiasm in the potential of scientific inquiry: To make learners have curiosity and imagination in the natural sciences
2. To build basic scientific literacy: To make learners equipped with basic scientific knowledge, inquiring spirit, and hands-on ability and to be able to communicate effectively, to be able to participate in public discussions and decision-making in social issues, to be able to reflect on science-related information reported in mass media, and to be able to develop the ability and habit of minds to seek the truth
3. To learn science and the applications of technology as lifelong practice: In order to be a modern citizen in an era of technology, it is necessary to develop learners' positive attitudes and interests in learning science
4. To develop the values of care and environmental protection and ability to take action
5. To prepare for career development: To promote learners' scientific literacy and prepare for their future development

In the draft of the Curriculum Guideline of Science Learning Area of 12-year Compulsory Education document, the key dimensions of science learning include core scientific concepts, scientific inquiry abilities (thinking ability and problem-solving), and attitude toward science and nature of science (three dimensions: interest of science, habits of scientific thinking and inquiry, and nature of science). Among the three dimensions, core scientific concepts are recognized as the "learning content" of science learning, while scientific inquiry ability and attitude toward science and nature of science are recognized as the desired "learning performance" of science learners. The implementation of science education should integrate the three dimensions according to learners' development and needs at different academic levels.

5.4.2 Development of Science Textbooks

Since the 9-year Compulsory Education Act was implemented in 1968, textbook development in Taiwan has been through four stages (Peng and Jan 2011):

1. Unified textbooks (1968–1989): In this period, all textbooks were edited, published, and disseminated by the National Institute for Compilation and Translation (NICT).
2. Unified and vetted textbooks (1989–2000): Most of the textbooks were published by NICT; some publishers were allowed to produce textbooks but still need to be vetted by NICT.
3. Vetted textbooks (2000–2005): Textbooks were open to free market competition. Book publishers were responsible for the production of textbooks, which were all vetted by NICT.
4. Vetted and MOE-sponsored (2005 to present): After the market for textbooks had been opened, the public realized some negative aspects, such as the rise in price or the quality issue. As a result, the MOE of Taiwan decided to publish MOE-sponsored versions of textbooks (in Mathematics Learning Area for 1–9 grades and Science Technology Learning Area for 7–9 graders) alongside the vetted textbooks.

After the release of the Grades 1–9 Curriculum Guidelines in 2000, the textbook market was opened up, and all book publishers were allowed to publish their science textbooks under the vetting of NICT. Some expert primary science teachers who had participated in editing the unified textbooks were recruited to edit science textbooks for publishers. At the same time, the number of expert primary science teachers was insufficient, and some teachers with less experience and scientists were recruited to edit primary science textbooks. As a result, the quality control of science textbooks has been a major concern for the government in Taiwan (Chiu 2007).

Since inquiry is the core characteristic of science education, inquiry-based instruction is advocated in primary science education in Taiwan. Primary science teachers in Taiwan are also expected to design suitable inquiry-based learning activities for their students by themselves. However, most primary science teachers rely heavily on textbooks and often adopt lecturing as the main instructional method as well as have their students conduct cookbook experiments. This may limit the scope of students' learning and reduce the opportunities for developing inquiry ability and understanding of nature of science.

5.5 Focus of Current Primary Science Teaching in Taiwan

5.5.1 Inquiry and Hands-On Learning Activities

Inquiry is the core of modern science education as was described in the National Science Education Standards (NSES) published in 2000 by the US National Research Council (NRC 2000). The NRC has also published a reference book titled *Inquiry and National Science Education Standards: A Guide for Teaching and Learning* for science teachers who want to carry out inquiry teaching. Influenced by the emphases of science education in America, the MOE of Taiwan has also emphasized the importance of inquiry learning activities. The Science Education White Paper published in 2003 stated that science education helps students acquire scientific knowledge and process skills and develop the habit of scientific thinking in scientific inquiry.

Currently, inquiry-based instruction is highly recommended in primary science in Taiwan. To promote such teaching, teachers are encouraged to share their lesson plans and instructional designs with others. We hope that through inquiry learning activities, science learners will have the ability to explore and argue, use scientific knowledge and process skills to solve problems, have a better understanding of nature of science, and finally build up positive attitudes toward science. In addition, according to the Grades 1–9 Curriculum Guidelines in Nature and Technology Learning Area, science education should be conducted using inquiry-oriented learning activities, which emphasize both hands-on and minds-on activities. However, assessment in Taiwanese primary and secondary schools relies mainly on paper-and-pencil tests, which overemphasize students' language literacy and logical thinking skills, rather than procedural knowledge. In view of this, the MOE of Taiwan has started holding Primary and Secondary Science Summer Camps for students in Grades 5–9 to enhance their hands-on ability to conduct experiments in 2014.

Many science educators have pointed out that doing science projects is the most common open inquiry activity in science classes (Huang 1999; Bell et al. 2005; Abd-El-Khalick et al. 2004). It was also reported that doing science fair projects can help learners construct deeper scientific knowledge, promote their conceptual understanding, help them understand the nature of science, improve their process skills, and develop their positive attitudes toward science (Bencze and Bowen 2009; Sumarll and Schillinger 2004). It has been over 50 years since the first national science fair projects were held in 1956 in Taiwan. Many primary and secondary teachers and students in Taiwan have participated in the annual science fair projects. The science fair projects in Taiwan are divided into three groups, namely, primary (Grades 1–6), junior high (Grades 7–9), and senior high (Grades 10–12). The purposes of the science fair project are to stimulate the interest in science learning and the potential of independent research, to enhance scientific abilities of thinking, creating, and innovating, and to cultivate students' correct attitudes toward science. However, in recent years, a lower quality of the science fair projects conducted by primary school students has been observed in Taiwan. This phenomenon revealed that Taiwan science teachers may face a lot of challenges, such as lack of professional knowledge, time, resources, and assistance, when conducting science fair instruction (Anderson 2002). Only a few science teachers know how to guide students to conduct science fair projects or inquiry activities effectively (Justi and Gilbert 2002). In addition, many science teachers have never learned science by using inquiry activities (Windschitl 2004), which was the main factor causing the lower quality of science fair projects.

In order to help primary science teachers enhance their professional knowledge in science fair instruction, the MOE of Taiwan held a competition of inquiry-based instruction design in 2012 that encouraged teachers to design lessons promoting students' inquiry ability. The inquiry-based instruction design will enable students to investigate and analyze scientific issues, to argue and explain, and to share their findings and reflections with their peers. Hopefully, by adopting inquiry-based instruction designed by expert primary science teachers, science teachers can turn their classes into a place where students engage in inquiry activities like scientists do.

5.5.2 Science Reading

Recently, science reading is also highly advocated in primary schools in Taiwan. In the process of scientific inquiry, reading plays an important role (Krajcik and Sutherland 2010). Also, science reading is one of the potential ways for those who have left school to learn science in informal learning contexts. Therefore, the habits of reading science-related articles or reports should be cultivated within school contexts. As a result, science reading is viewed as one of the critical components of primary science education in Taiwan. Another reason why science reading is emphasized in Taiwan is the promotion of public science education. There are several channels for accessing public science in Taiwan, such as the media, newspapers, lectures, exhibitions, and relevant science activities held by civil societies. Among these learning channels, reading scientific articles is a direct and accessible exposure to science for the public. In 2010, the activity of reading articles that were written for the public was strongly promoted by the MOE in Taiwan. It is advocated that scientific knowledge is the foundation of critical thinking for citizens. By reading popular science articles, people will be more curious about science and can have reasonable levels of critical thinking ability. Since 2010, science reading materials as well as science reading activities have been introduced to primary schools in Taiwan. It is hoped that students' scientific literacy can be improved by science reading.

5.5.3 Outdoor Education

Moreover, outdoor education has been advocated in Taiwan in recent years. Outdoor education usually refers to organized learning which takes place in the outdoors, including school grounds or school environments, tribal/indigenous areas, forests, nature, and cultural exchange activities. In traditional school contexts, learners are trapped in the classroom and are far away from the outer world. Therefore, the 2013 Taiwan outdoor education declaration issued by MOE in Taiwan especially emphasized the importance of outdoor education. By promoting outdoor learning, learners can cultivate ethics, attitudes, and creativity for cultivating independence, strengthen motivation for learning, and increase cognitive development (MOE 2015). Based on the declaration, Taiwan held an "Outdoor Education Summit Conference" in 2016, and parents, teachers, and school administrators were expected to discuss their vision of outdoor education in Taiwan.

5.5.4 Integrating Maker Education into Primary Science Education

For many years, the significance of STEM education has been recognized by science educators in Taiwan. In fact, it is still very difficult to implement STEM education in primary science classes in Taiwan. However, with the integration of maker education, it offers much promise for STEM education efforts in Taiwan.

Maker education has received greater attention in Taiwanese primary schools since 2015 in two major focal areas: One emphasizes the cultivation of computational thinking while the other uses 3D printers to design and make artifacts. Regarding the former, students are expected to learn designing and coding with some basic coding languages such as Scratch. Then they may proceed to use open-source electronic prototyping platforms (e.g., Arduino) to create interactive electronic objects or robotics to solve problems. Regarding the latter focus, it is more related to engineering modeling. In engineering, modeling is very important in design process. With thc use of 3D printers, it is much easier for students to turn their own design into concrete artifacts, and then they could test them and redesign iteratively in modeling stage of engineering design process. However, in Taiwanese primary schools, 3D printers function as typical machines that "print out something designed by others" rather than produce novel artifacts that have been designed by students themselves.

In recent years, many workshops related to maker education have been organized for primary science teachers in Taiwan. Teachers are expected to learn about the two major aims of maker education and are encouraged to integrate maker education into their science classes. However, due to lack of prior knowledge, skill, and experiences of maker education, most primary science teachers have difficulty integrating maker education or engineering education into their science teaching. It seems that more well-organized and long-term professional development programs on maker or engineering education are needed. Due to the shortage of expert teachers and the restricted resources for maker education (such as high cost of robotics and 3D printers), not every student can involve in maker education.

In sum, maker education could be one of the important foundations for implementing STEM education. Nowadays, more and more primary science teachers in Taiwan increasingly recognize its importance and try to implement this in their classes. However, it may be still a big challenge for them to integrate what they learned about maker education into their science classes. In order to spread maker education, more teacher professional development and investment of maker education resources should be highlighted and promoted by the education authorities in Taiwan.

5.6 Primary Science Teacher Education and Professional Development

5.6.1 Primary Science Teacher Education

Before 1994, primary teachers (including science teachers in general) in Taiwan were educated at public expense. A total of nine teacher colleges were the main institutions devoted to teacher education. In particular, only the students from the departments of mathematics and science education in these nine teacher colleges were trained to be primary science teachers. Other students in teacher colleges had to take basic science courses with teaching methods in science courses. All students who graduated from these teacher colleges were viewed as qualified teachers to teach any subject in elementary schools.

At that time, due to administrative considerations, primary science courses were not necessarily taught by teachers who had graduated from the department of mathematics/science education. Even though, at that time, most elementary teachers did not graduate from the department of mathematics and science education, they had basic scientific knowledge and received sufficient training in teaching primary science, and they were able to teach primary science.

Since the Teacher Education Act was announced in 1994, teacher education programs were not confined to only universities of education (former teacher colleges). All universities could now apply to set up teacher education centers and primary teacher education programs in which students can become primary teachers as long as they obtained the required credits (see Table 5.1). However, by adopting this teacher education approach, there were no dedicated departments for preparing primary science teachers, and the number of primary teachers with science-related backgrounds decreased.

It is noteworthy that, unlike those graduates from teacher colleges before 1994, after the launch of Teacher Education Act, pre-service primary teachers are not

Table 5.1 Comparison of the two primary teacher education systems in Taiwan

	Before 1994	After 1994
The department of educating science teachers	Yes (department of mathematics and science education in Teachers College)	None
Required science education courses for pre-service primary teachers	"Basic science courses," "introduction to technology," and "instructional methods and learning materials of primary science"	None
Internship of pre-service teachers	One full semester year after graduating	The last semester of the teacher education program
The identity of pre-service teachers	Student teachers	Students
Internship in science teaching	Not required	Not required
Certificate of science teachers	None	None

strictly required to take relevant science courses such as "basic science courses," "introduction to technology," or "instructional methods and learning materials of primary science." This has resulted in many primary teachers not having sufficient pedagogical content knowledge (PCK) in teaching primary science and may directly have adverse influences on primary school students' science learning.

Another problem with primary science teacher education in Taiwan is the process of the internship of pre-service teachers. During the internship, practicing teachers who guide pre-service teachers, in general, emphasize classroom management skills and the two major courses, mathematics and Chinese. Thus, many pre-service teachers do not have any opportunities in practicing teaching primary science (Lian and Hsiung 2010).

5.6.2 *Primary Science Teacher Professional Development*

Undoubtedly, teachers should be regarded as professionals (Blackman 1989; Nodding 1992). Thus, like other professionals (such as doctors, accountants), teachers also need to have continuous professional development. According to the current rules from the MOE, primary teachers need to attend further studies for 18 h every semester year or accumulate 90 h (5 credits) in a 5-year period. In Taiwan, Wednesday afternoon is a fixed period of time for in-service primary teachers' professional development, and workshops held in this period of time are the main channels for in-service primary teachers' professional development. Usually, primary schools arrange professional development workshops for every subject area on Wednesday afternoons, but most of the workshops are for language and mathematics, and science education workshops only account for just a small part.

In addition to workshops, serving teachers are also encouraged to take professional training such as master's or doctoral degrees. According to the survey conducted by the Ministry of Education in 2013, almost 40% of primary teachers in Taiwan hold a master degree. However, not all primary science teachers hold a degree in science or science education.

As the Grades 1–9 Curriculum has recommended, all local governments have established mentor groups for primary science education which are responsible for helping serving science teachers develop their professional abilities and improve their skills in cultivating students' essential competences. With open invitations, excellent and experienced primary science teachers are selected as members of the mentor groups for primary science education. These mentor groups for primary science education are responsible for helping younger teachers improve their instructional design and giving guidance on novice science teachers' science instruction. They also assist schools to develop school-based science curricula and creative/innovative instruction (Ling and Chen 2004). The mentor groups are also responsible for holding workshops in science education and giving science teaching demonstrations.

In 2007, the MOE of Taiwan proposed an integrated program which aimed to integrate different hierarchical educational organizations, including municipal, city, and county governments, national education mentor groups, and schools. In recent years, Taiwanese science mentor groups have pushed forward the national science fair and carried out the development of elementary school science as a field and teacher professional development by promoting science reading, inquiry teaching, teaching assessment, extracurricular activities, and science education affairs. To this end, these mentors had to design science learning activities related to science reading, inquiry learning, assessment, and extracurricular activities. In each year, these mentors have to select some schools for conducting science education supervision and guidance. During the supervision and guidance in selected schools, they often demonstrate science learning activities they designed and discuss with science teachers in these schools. These mentors are also invited as lecturers of the Wednesday afternoon workshops for teachers' professional development.

The MOE of Taiwan also holds annual fairs of science education aiming to create a platform for national and local mentor groups of primary science education to share educational research and practices, to build up a means of sharing teaching strategies, to innovate teaching and curriculum, and finally to promote the interaction between the national mentor group and other local mentor groups in Taiwan. In the annual fairs of science education, science teaching materials, teaching aids, publications (such as reports and learning activities design), and research papers from all local member groups are presented. Science teachers' professional teaching abilities can be enhanced by dialogue with professional professors in theme-based sessions. All the participants therefore have a chance to interact with each other and learn from each other, and member groups can share and exchange their current science education resources.

Furthermore, more and more schools actively encourage the formation of teachers' professional development and lesson preparing communities. In addition to school teacher professional groups and lesson preparing communities, there are many science teachers' communities on *Facebook* where teachers have more opportunities to discuss and share ideas with one another. By experiencing diversified learning experiences online, primary science teachers in Taiwan can discuss, plan, explore, design, and reflect together for finding students' learning difficulties and thus enhance students' learning outcome.

The primary teachers' responses to the government's professionalization in science teaching may vary. Most of the primary science teachers are not science majors, and they may not have to teach science every year. For them, attending professional development workshops may just meet the government minimum requirement. But, for some primary science teachers, they teach science every year, and they enjoy their teaching. They are often eager to improve their profession in teaching primary science. As a result, they are more positive about the government's professionalization in science teaching.

5.6.3 Accreditation of Expertise in Science

In Taiwan, most primary teachers have to teach more than one subject. That is, non-science majors in primary schools also need to teach science. In practice, almost all the primary schools here have science teachers that do not have sufficient professional knowledge in teaching science because for a long time, the upgrading of non-science majors in primary science has been neglected. In order to promote the quality of educational experiences, the MOE has mandated that elementary teachers who are non-science majors need to have the accreditation of science.

For primary school teachers, to get the necessary accreditation in science, they should have sufficient science domain knowledge and professional ability. In order to carry out the examining and verifying process, the MOE of Taiwan assigned this important task to National Taichung University of Education (NTCU) in 2012. The NTCU has invited different area experts to manage the examining and verifying process for supervising primary school teachers' expertise accreditation in science.

The approach of getting the accreditation in science teaching is listed below (see Table 5.2). It is estimated that all current primary teachers who teach science need to have this expert accreditation in science teaching. By doing this, it is believed that the shortage of primary science teachers in Taiwan can be solved in the long term.

5.6.4 Primary Science Teacher Research

To promote science education in Grades 1–9, the MOE of Taiwan actively runs the "Primary and Secondary Science Education Research Project" which encourages science teachers to conduct science education research. It is hoped that the quality

Table 5.2 Approaches of getting expertise accreditation in elementary science teaching

Identity		Approach
Pre-service teachers	Pre-service teachers	Successful completion of science domain knowledge modules worth 22 credits and pass the science domain knowledge assessment
Practicing teachers	Primary teachers who do not teach science	Successful completion of science domain knowledge modules worth 22 credits and pass the science domain knowledge assessment
	Current science teachers (years of service <1)	Successful completion of science domain knowledge modules worth 22 credits and pass the science domain knowledge assessment
	Current science teachers (1 < years of service <5)	Successful completion of science domain knowledge modules worth 4 credits or participate in 72 h of workshops and pass the science domain knowledge assessment
	Current science teachers (years of service >5)	Pass the science domain knowledge assessment

of primary and secondary science education can be enhanced through the practice of such projects. The main topics in these research proposals include:

1. Environmental science education activities: The activities include environmental science winter and summer camps, experimental or exemplary education activities, environmental science education publications or websites, and others.
2. The research and development of science teaching materials, methods, and assessment: Research of teaching materials, assessment in nature and technology subjects, digitalized information, and science courses.
3. The research and guidance of talented science students: This includes the research on teaching and assessment, revision of remedial/supplementary materials, the conduct of educational activities, and follow-up guidance of talented science students.
4. The research and promotion of local science teaching materials: Including the research of local science teaching resources and materials, the promotion of supplementary materials, and the environmental education activities for primary and secondary students.
5. The research and development of creative science activities: The conduct of primary and secondary science activities and the research on effective materials.

In recent years, many primary science teachers prefer conducting action research to solve their own practical teaching problems under the circumstances of curriculum reform. To promote curriculum development in the subject, teachers have been very active in conducting action research and are willing to share their results with other teachers. For example, Chen (2011) reported a three-phase action research regarding a teacher's professional development in designing and implementing science reading activities in her classroom. Various qualitative and quantitative data from the teacher, the students, a teacher educator, and classroom teaching were collected and analyzed. The teacher reflected on the conduct of all the three rounds of science reading activities, and her reflections were focused on her beliefs, instructional design, and teaching practices regarding science reading activities, as well as on the conduct of teacher action research. Then, the teacher summarized some practical advice about the implementation of science reading activities in the elementary classrooms. Chou (2012) is another example of action research; this was a female science teacher's progression regarding integrating socio-scientific issues (SSIs) into science instruction, including designing learning activities, teaching practices, and assessment practices. Because of the action research, the teacher's pedagogical content knowledge enhanced during the period of the SSI-based instruction. Based on her self-reflection on the action research, the teacher also provided suggestions for teachers who are interested in SSI-based instruction.

5.6.5 Primary Science Education Research Published in the Local Science Education Journal

In Taiwan, the *Chinese Journal of Science Education* (CJSE) published since 1993 is the official journal of the Association of Science Education in Taiwan. Due to its wide scope and contribution in Taiwan's science education, it has also been accepted in 2008 by Taiwan Social Sciences Citation Index (TSSCI). In order to provide the overview of primary science education research in Taiwan, the authors have conducted a content analysis of CJSE publications related to primary science education in Taiwan. Among the science education research papers published in CJSE from 2001 to 2015, a total of 79 studies focusing on primary science education research were selected and analyzed in terms of research topics, research contexts, sample groups, and research methods.

For the research topics, we included eight categories adapted from the National Association for the Research in Science Teaching (NARST) Conference strand categories. The eight categories are:

1. Teacher education: including pre-service and continuing professional development of teachers
2. Teaching: including teacher cognition, pedagogical knowledge and pedagogical content knowledge, teacher thinking, and teacher behaviors and strategies
3. Learning conceptions: including students' conceptions and conceptual change, conceptual change in learners, and conceptual development
4. Learning contexts: including classroom contexts and learner characteristics, student motivation, learning environment, individual differences, reasoning, learning approaches, exceptionality, teacher-student interactions, peer interactions, laboratory environments, affective dimensions of science learning, cooperative learning, language, writing and discourse in learning, and social, political, and economic factors

5. Goals and policy, evaluation, and assessment: including curriculum development, evaluation, assessment, and curriculum policy and reform
6. Culture, social, and gender issues: including multicultural and bilingual issues, ethnic issues, and issues related to science teaching and learning
7. History, philosophy, epistemology, and nature of science: including historical issues, philosophical issues, epistemological issues, ethical and moral issues, and nature of science
8. Educational technology: including computers, interactive multimedia, video, integration of technology into teaching, and learning and assessment involving the use of technology (Tsai and Wen 2005).

From 2001 to 2015, the research topics of primary science education research in Taiwan were mainly learning conceptions (28%) and learning contexts (28%), followed by teaching (22%) and goals and policy, evaluation, and assessment (14%). Undoubtedly, conceptual learning is always an important issue in primary science education research in Taiwan. However, primary science education researchers in Taiwan have also highlighted students' reasoning and argumentation and higher-order thinking skills. It is noteworthy that there were few studies focusing on teacher education (2.5%); history, philosophy, epistemology and nature of science (3.8%); educational technology (1.3%); and culture, social, and gender issues (1.3%). In recent years, Taiwan has strongly promoted digital learning in primary schools, and primary science teachers have tried to use educational technology to support their teaching. It is noted that most of the relevant studies were published in educational technology journals but only a few relevant studies were published in the CJSE.

Regarding the research contexts, we categorized the research contexts in the selected studies into three major types: formal education for students, informal education for students, and professional development for teachers. The studies focusing on primary science education were mainly conducted in formal educational contexts (51.9%), followed by informal educational contexts (27.8%), and professional development for teachers (20.3%), revealing that the primary science education research in Taiwan were mostly undertaken in school contexts. Nevertheless, it should be noticed that informal science education is still highly highlighted in the primary science education research in Taiwan.

As for sample groups, students and teachers were identified, respectively. Meanwhile, the research methods were recognized. Among the sample groups analyzed from 2001 to 2015, the distribution percentage of K5–6 is the highest (57.3%), followed by K3–4 (28.4%), in-service teachers (18.2%), K1–2 (5.7%), and pre-service teachers (2.3%). The results show that K5–6 students was the most frequently used sample group from 2001 to 2015, and there was few research focusing on pre-service teachers (2.3%) and K1–2 (5.7%). It suggests that future research could pay more attention on K1–2 and pre-service primary science teachers.

Finally, among the research papers published in the CJSE during 2001–2015, the more frequently used research methods in the primary science education in Taiwan are mixed methods (58.2%), followed by qualitative methods (22.9%) and quantitative methods (19%). The results indicate that more relative studies in primary science education research in Taiwan tended to use both qualitative and quantitative methods.

5.7 Achievement Assessment in Primary Science Education

5.7.1 Assessment in Schools

Traditionally, summative assessment is used to evaluate primary students' science achievements in Taiwan. Normally, students' achievements in science education are assessed by paper-and-pencil tests. Due to the fact that most of primary science teachers heavily rely on textbooks, the items in the paper-and-pencil tests are designed to align with the content covered in the textbooks. Most of these paper-and-pencil tests are standardized tests and require one correct answer on the items. These tests mostly focus on assessing students' conceptual understanding, rather than their process skills, inquiry ability, or understanding of the nature of science. Due to rudimentary or inadequate backgrounds in science education, most primary science teachers are not familiar with contemporary laboratory or hands-on instruction. Therefore, it is a huge challenge for science teachers to adopt assessments to evaluate students' hands-on or process skills as well as inquiry ability.

In recent years, Taiwanese primary science teachers are encouraged to utilize formative assessment and portfolio assessment for evaluating students' science learning. In formative assessment, teachers conduct in-process evaluations of student comprehension as well as learning needs and progress during an instructional lesson or unit. By collecting various data regarding student learning process and outcomes, portfolio assessment enables the measurement of multiple dimensions of student learning progress (Venn 2000). However, it requires extra time to plan an assessment system and conduct the assessment in both formative assessment and portfolio assessment. Also, formative assessment and portfolio assessment involve the extensive use of subjective evaluation procedures, such as rating scales and professional judgment, and this limits reliability. Therefore, still not many primary science teachers in Taiwan use these two assessment methods in evaluating students' science learning.

5.7.2 Nationwide Assessment of Student Achievement in Taiwan

In 2005, a newly nationwide assessment study, Taiwan Assessment of Student Achievement (TASA), was launched (http://tasa.naer.edu.tw/15news-1.asp). Before the launch of TASA, there are four major international student achievement assessments that Taiwan students participated in: the Trends in International Mathematics and Science Study (TIMSS), National Assessment of Educational Progress (NAEP), Programme for International Student Assessment (PISA), and Progress of International Reading Literacy Study (PIRLS). However, there was no database for continuously collecting data of student achievement in primary, middle, and high schools. The aim of the TASA data collection is to examine Taiwanese student performance on language (Chinese and English), science, mathematics, and social studies in primary, middle, and high schools. At the elementary level, the target students of TASA are fourth and sixth graders.

The collected data in TASA database could be used to conduct standardized analyses of student science achievement. Also, it could provide information on the trends of changes in students' ability at different academic levels, which could be helpful for understanding the effectiveness of the implemented science curriculum and the modification of science curriculum. Most importantly, the launch of TASA study is aligned with international interest in assessing student science achievement at a state level. Thus, it is helpful for comparing Taiwanese students' science achievement with those from other countries in the world, providing insights into the unique as well as the strength and the weakness of science education in Taiwan.

The TASA assessment items in science achievement for K4-6 graders are constructed based on the competence indicators listed in Grades 1–9 Curriculum Guidelines in Nature and Technology Learning Area for K4-6 graders. The scope of TASA assessment items in science achievement for primary school students covers:

1. Process skills, including (1) observation (students can observe that objects, organisms, and events have recognizable characteristics and attributes; can categorize and classify objects, organisms, and events according to their attributes), (2) induction and inference (students can summarize data collected from experiments and propose research findings), and (3) interpreting data (students can develop explanations for phenomena or predict what will happen etc.)
2. Knowledge in science and technology, including knowledge about animals, plants, materials, and environments (weather, moon)
3. Nature of science, such as understanding that ideas can be tested by conducting experiments, etc.
4. Application of science: can use scientific knowledge or results of experiment to solve a problem

The TASA is an online assessment system with two types of items in science for elementary school students. One is multiple-choice item which has one "correct" or "better" answer. The other is matching item. The data collected in the TASA is available to the public as open educational data.

5.7.3 International Assessment Student Achievement in Taiwan

Both TIMMS and PISA are international assessments on students' mathematics and science achievement. Specifically, TIMMS administered in four-year cycles assesses primary school students' (fourth graders) science achievements. In 2011, selected students from 63 countries including Taiwan participated in TIMSS. The TIMSS science assessment is based on a comprehensive framework including two dimensions: content dimension and cognitive dimension. The content dimension is concerned with the domains or subject matter to be assessed within science, which include life science (45%), physical science (35%), and earth science (20%). The cognitive dimension is concerned with the domains or thinking processes expected of students as they engage with the science content, which include knowing, applying, and reasoning.

In TIMSS 2011, Taiwan is the fifth top-performing countries (the same as the Russian Federation), after Korea, Singapore, Finland, and Japan, in science in TIMSS 2011 at the fourth grade. In 2003, 2007, and 2011, Taiwan remained steady in TIMSS results. In TIMSS 2011, the fourth graders in Taiwan significantly scored higher than the TIMSS average scores in the three content domains (i.e., life science, physical science, and earth science) and the three cognitive domains (i.e., knowing, applying, and reasoning). In all the content and cognitive domains, Taiwan is one of the ten top-performing countries. Regarding gender difference, the male fourth graders significantly outperformed the females in their overall science achievement in TIMSS 2011.

Moreover, 58% of the fourth graders in Taiwan expressed positive attitudes in learning science, 30% expressed slightly positive attitudes in learning science, and 11% expressed negative attitudes in learning science in TIMSS 2011, revealing that in general the fourth graders in Taiwan expressed relatively positive attitudes in learning science. Regarding confidence in learning science, 44% of the fourth graders in Taiwan expressed that they were confident in learning science, 35% expressed that they had little confidence in learning science, and 21% expressed that they had no confidence in learning science. It should be noticed that although Taiwan is one of the ten top performing countries listed in TIMSS 2011, the interest and the confidence in learning science of fourth graders are not within the top ten-performing countries.

Lu and Lien (2016) compared Taiwan's TIMSS 2011 performances with those of other top-ranking countries. The findings showed that Taiwan's elementary science education emphasized experiments and physical science was the emphasized content in the fourth grade textbooks. Also, they found the relative low correlation between Taiwan's curriculum and TIMSS examination, science class hours in Taiwan are relatively lower, and Taiwan's elementary science education has few emphases on "reading," "memorizing," and "explanation." The findings above provide some implications for future innovations in primary science education in Taiwan. In particular, primary science learning in Taiwan may need to be more learner-centered; for example, we should provide students with more opportunities to engage in active reading, constructing, organizing, and integrating their own ideas about and of science, formulating their own scientific explanations during their hands-on activities, and sharing and communicating their own ideas with peers. That is, the implementation and increase of minds-on learning activities as well as integrating minds-on and hands-on learning activities could be crucial issues in primary science education in Taiwan.

5.8 Concluding Remarks

This chapter focuses on the evolution and the current status of primary science education in Taiwan. In this chapter, an overview of science education in Taiwan is introduced. To respond to the huge challenges of national competence in the knowledge economy era, the educational authority in Taiwan is heavily committed to preparing scientifically and technically literate citizens (Lu and Lien 2016).

As for the primary science curriculum and textbooks, a new curriculum guideline is now being prepared in Taiwan. How the new science curriculum guideline can help to cultivate national scientific literacy in Taiwan should be an important issue for primary science educators and teachers. Moreover, in Taiwan, the science textbooks are no longer controlled by the authority. And most primary science teachers in Taiwan rely heavily on science textbooks. Therefore, the quality of primary science textbooks is a crucial issue worth addressing by science educators and the government.

Primary science education plays an important role in cultivating national scientific literacy. Hence, teacher education or professional development for primary science teachers is crucial. Regarding the science teacher education system, after some major changes, primary teachers are now required to have science expertise accreditation in 2018, which may promote primary teachers' professional ability in teaching science and enhance students' science learning outcome. It seems an effective solution to the low quality of primary science instruction. Meanwhile, the Ministry of Education of Taiwan should rethink about how to prepare primary teachers' professional ability in teacher education programs or whether primary science courses should be taught by teachers with science-related backgrounds.

For a long time, Taiwanese primary students have shown high science achievement performance in TIMSS. However, in TIMSS 2011, although Taiwan is one of the ten top-performing countries, the interest and the confidence in learning science of fourth graders are not within the top 10 performing countries. It will be one of the significant issues for primary science educators and teachers. One of the possible reasons may be that, normally, primary science courses are lecture-based and primary students' achievements in science are majorly assessed by paper-and-pencil tests in Taiwan. Through inquiry learning activities, science learners can have the opportunities to do hands-on activities, and their attitudes and confidence in learning science may be improved. Moreover, for primary science teachers, they have to develop their professional ability in designing and implementing inquiry-based instruction and hands-on learning activities. They also have to get familiar with alternative assessments, rather than summative assessments as pencil-paper tests.

For primary science education in Taiwan, now it is a crisis but also a turning point. To improve primary science education in Taiwan, the MOE, various educational authorities, science educators, schools, and science teachers have to work collaboratively. Thus, the quality of primary science education in Taiwan could be improved, and our national scientific literacy could be promoted to be one of the top rank countries in the world.

Acknowledgments Funding of this research work was supported by the Ministry of Science and Technology, Taiwan, under grant numbers MOST 103-2511-S-008 -007 -MY3 and MOST 104-2511-S-008 -015 -MY3. The authors would like to express their gratitude to the editor and reviewers for their helpful comments in the further development of this book chapter.

References

Abd-El-Khalick, F., BouJaoude, S., Duschl, R., Lederman, N. G., Mamlok-Naaman, R., & Hofstein, A. (2004). Inquiry in science education: International perspectives. *Science Education, 88*(3), 397–419.

Anderson, R. D. (2002). Reforming science teaching: What research says about inquiry. *Journal of Science Teacher Education, 13*(1), 1–12.

Bell, R. L., Smetana, L., & Binns, I. (2005). Simplifying inquiry instruction. *The Science Teacher, 2005*, 30–33.

Bencze, J. L., & Bowen, G. M. (2009). A National Science Fair: Exhibiting support for the knowledge economy. *International Journal of Science Education, 31*(18), 2459–2483.

Blackman, C. A. (1989). Issue in professional development: The continuing agenda. In M. L. Holly & C. S. McLoughlin (Eds.), *Perspective on teacher professional development* (pp. 1–15). New York: The Falmer Press.

Chen, Y.-P. (2011). *An action research on implementing science reading activities for fifth and sixth graders* (Unpublished master thesis). National Taichung Normal University, Taiwan, ROC.

Cheng, Y.-J. (2005). Review and outlook of science education reform in Taiwan. *Science Education Monthly, 284*, 2–22 (in Chinese).

Chiu, M. H. (2007). Standards for science education in Taiwan. In D. J. Waddington, P. Nentwig, & S. Schanze (Eds.), *Making it comparable: Standards in science education* (pp. 303–346). Münster: Waxmann.

Chou, P.-C. (2012). *An action research on implementing ICT-enhanced SSI-based instruction in a remote-district elementary school* (Unpublished master thesis). National Taichung Normal University, Taiwan, ROC.

Huang, H.-B. (1999). *Research of student-centered scientific inquiry activities*. MOE report of primary and secondary science education funding.

Justi, R. S., & Gilbert, J. K. (2002). Science teachers' knowledge about and attitudes towards the use of models and modeling in learning science. *International Journal of Science Education, 24*(12), 1273–1292.

Krajcik, J. S., & Sutherland, L. M. (2010). Supporting students in developing literacy in science. *Science, 328*(5977), 456–459.

Lian, S.-H., & Hsiung, C.-T. (2010). A study of developing elementary pre-service science Teachers' pedagogical content knowledge by teacher education institution and elementary schools. *Research and Development in Science Education Quarterly, 57*, 21–54 (in Chinese).
Ling, K.-L., & Chen, C.-M. (2004). Grade 1–9 curriculum reform for city and county level. *Education Research Monthly, 122*, 47–59 (in Chinese).
Lu, Y. L., & Lien, C. J. (2016). Elementary science education in Taiwan—From the perspective of international comparison. In *Science education research and practices in Taiwan* (pp. 163–180). Singapore: Springer.
Ministry of Education. (2015). *National Guidelines for outdoor education*. Taipei: MOE.
Ministry of Education and National Science Council. (2003a). *The white paper for science education*. Taipei: MOE.
Ministry of Education and National Science Council. (2003b). *Grade 1–9 curriculum guidelines: Science and technology learning area*. Taipei: MOE.
National Research Council (NRC). (2000). *Inquiry in the national science education standards: A guide for teaching and learning*. Washington, DC: National Academy Press.
Nodding, N. (1992). Professionalization and mathematics teaching. In D. A. Groups (Ed.), *Handbook of research on mathematics teaching and learning: A project of national council of teachers of mathematics* (pp. 197–208). New York: Macmillan.
Sumrall, W., & Schillinger, D. (2004). Non-traditional characteristics of a successful science fair project. *Science Scope, 27*(6), 20–24.
Peng, C.-L., & Jan, M.-H. (2011). An analysis of the controversy over the policy of concurrently reviewing and approving publicly and privately compiled textbooks. *Journal of Textbook Research, 4*(1), 1–30 (in Chinese).
Tsai, C.-C., & Wen, L. M. C. (2005). Research and trends in science education from 1998 to 2002: A content analysis of publication in selected journals. *International Journal of Science Education, 27*, 3–14.
Windschitl, M. (2004). Folk theories of "inquiry:" How preservice teachers reproduce the discourse and practices of a theoretical scientific method. *Journal of Research in Science Teaching, 41*(5), 481–512.

Chapter 6
Progress and Challenges of Elementary Science Education in Korea

Mijung Kim, Hye-Gyoung Yoon, and Mee-Kyeong Lee

6.1 Introduction

Science and technology have been regarded as priorities to enhance national sovereignty, economy, and international relationships in Korea. There have been various efforts to promote science and technology to develop the nation since the aftermath of the Korean War during 1950–1953. The focus on innovations in science and technology and Research and Development (R&D) industries has continued in Korea to survive and prosper in the globalized world (OECD 2007, 2014). Under current economic uncertainties and labor market demands, the emphasis of scientific and technological innovations and problem solving are evident as well as necessary such as the recent emphases on STEM (Science, Technology, Engineering and Mathematics) literacy (Ministry of Education Science and Technology 2011) in the country. There are also concerns about new challenges and ethical dilemmas in society such as the energy crisis, global warming, climate change, ethics in biotechnology, and so forth. (Hodson 2003; Kolsto 2001; Osborne 2010). These expand the understanding of scientific literacy from just conceptual knowledge and practical skills towards building citizenry responsibilities of science in society.

Students' critical thinking, decision making and problem solving in everyday and social problems now became key components of scientific literacy. With the

M. Kim (✉)
University of Alberta, Edmonton, Canada
e-mail: mijung@ualberta.ca

H.-G. Yoon
Chuncheon National University of Education, Chuncheon, South Korea
e-mail: yoonhk@cnue.ac.kr

M.-K. Lee
Korea Institute for Curriculum and Evaluation, Chungcheongbuk-do, South Korea
e-mail: mklee@kice.re.kr

Y.-J. Lee, J. Tan (eds.), *Primary Science Education in East Asia*, Contemporary Trends and Issues in Science Education 47,
https://doi.org/10.1007/978-3-319-97167-4_6

focus of scientific literacy for both the future economy and citizenship, the Korean government strives to enhance science education in schools through developing an effective science curriculum, teaching resources and materials, assessment strategies, enhancing classroom environments, and reforming educational policies. Since the first science curriculum in 1940s, there has been much progress in terms of clarifying the rationales and practices of science education nationwide. In this chapter, we introduce how science as one of the core courses in the Korean education system has been developed, supported, and practiced in local schools and classrooms. We then give an overview of the Korean education system in national and international contexts, students' achievements in international assessments such as the Trends in International Mathematics and Science Study (TIMSS), the development of the elementary science curriculum, teacher education programs, assessment practices, and some pedagogical issues and concerns in elementary science education in Korea. In addition, we will discuss concerns and challenges for further development of science education in Korea.

6.2 Overview of Elementary Science Education in Korea

Korea has more than 5000 years of history and there have been various types of schools over that long history. The first educational instituition was *Taehak* (태학, 太學), established in 372 AD by the Goguryeo, one of the Three Kingdoms of Korea. The curriculum consisted of ethics education focused on cultivating the morals of students and educating the public based on Confucianism and Buddhism. Modern schools similar to Western models were introduced in the 1880s and they consisted of national and private educational institutions established by Christian missionaries and members of the independence movement. The Korean government established a modern educational system in 1948 after Korea's independence from Japan. Elementary education became compulsory in 1953 and middle school education in 1985 (Korea Institute of Curriculum & Evaluation 2012). The fact that elementary education became compulsory in 1953 showed that Korea recognized and prioritized the importance of education despite the country's extreme poverty in the wake of Japanese colonialism and the Korean War. Education has arguably been the most important factor in Korea's rapid growth and development; Korea's only natural resources are its people. Most Koreans have therefore sought education as the most effective way to improve their social status and the country is now famous for its passion for education—the so-called educational fever or educational zeal (Seth 2012). Since the establishment of modern educational systems, Korea has a single-track 6–3–3–4 system: six years for elementary school, three for middle school, three for high school, and four years of tertiary education. Students enter elementary school at age of seven, after one or two years of kindergarten. Most children attend a public school near their home, as allotted by the education board. Private elementary schools select students through their own lottery system.

In terms of the organization of educational administration in Korea, there are three layers of administrative authority; the national Ministry of Education, municipal and provincial Offices of Education, and county Offices of Education. These offices supervise elementary and middle schools. The Korean government has traditionally been very centralized, but since the reform initiatives in 1995 there have been efforts to decentralize the system and allow more decision-making autonomy at the regional, local, and school levels. From the beginning of 1996, a Schools Council has operated in every national or public elementary and secondary school to secure the autonomy of the school's management, and to maximize community participation. The Schools Council deliberates on school budgets, curriculum implementation, management of extra-curricular activities, cooperative activities between schools and community, and student welfare.

Korea has a national curriculum; the first national curriculum was established in 1954, a year after the end of the Korean War and has been revised 10 times since then. The new national curriculum was inaugurated in September 2015 and will be implemented in schools from 2017. We will discuss the details of the development of curriculum in later sections.

Elementary school education has focused on the cultivation of basic abilities, the formation of everyday habits needed for student learning and daily life, and the development of good character. The 2015 national curriculum states the following goals of elementary education (Ministry of Education, Korea 2015):

1. To know the value of oneself, to grow healthy life habits and to find and represent dreams through a variety of learning experiences.
2. To develop basic abilities to recognize and solve problems at school and in daily life and to cultivate the imagination to make new experiences.
3. To enjoy cultural activities and to create appropriate attitudes to experience happiness in nature and everyday lives.
4. To encourage students follow rules and orders, to sympathize and cooperate with others, and to cultivate the mind to be considerate of others.

Science is one of the school subjects in elementary schools and it is integrated with other subjects in Grades 1–2 and becomes a separate subject from Grade 3. We will explain the details of the components of elementary school subjects and science education in a later section, but we introduce the goals of elementary science curriculum here to understand how the goals of elementary education in general and science curriculum in particular are aligned and interconnected. The goals of the elementary science curriculum are as follows:

> The science curriculum aims to help students understand the basic concepts of science and develop inquiry skills with interest in and curiosity about natural phenomena and objects. In consequence, students develop scientific literacy necessary for solving creatively and scientifically personal and social problems in daily life. The objectives of the science curriculum are to educate students who are able to:
>
> 1. Enhance curiosity and interest in natural phenomena, and develop an attitude to scientifically solve problems in daily life.
> 2. Develop the ability to solve problems of natural phenomena and daily life scientifically.

3. Understand the basic concepts of science by doing inquiry about natural phenomena.
4. Recognize the relationship between science, technology and society and develop literacy as a citizen in a democracy.
5. Develop lifelong learning skills in recognition of joy of learning science and usefulness of science (Ministry of Education, Korea 2015).

As stated in the Korean national science curriculum, the ultimate goal of elementary science education is to develop scientific literacy. To achieve this goal, enhancing curiosity and interest in science is emphasized in addition to gaining scientific knowledge. In addition, engaging in science through inquiry activities is also important for developing scientific thinking and problem solving skills. Both the goal statements of elementary education and science curriculum recognize and emphasize the importance of enhancing positive attitudes toward learning, developing problem solving skills, connecting schools and everyday lifeworlds, and engaging students' daily experiences in learning.

6.3 Student Achievement in International Assessments: TIMSS Results from Korea

In this part of the chapter, we want to describe some results from student participation in international science achievement tests and how these findings have influenced curriculum development and other educational policies in Korea. The Trends in International Mathematics and Science Study (TIMSS) is a large-scale assessment designed to inform educational policy and practice by bringing an international perspective to teaching and learning in mathematics and science. TIMSS was established by the International Association for the Evaluation of Educational Achievement (IEA). TIMSS assesses achievement in mathematics and science of students in Grades 4 and 8 and collects background information to address concerns about school resources and the quality of school curriculum and instruction. TIMSS has been conducted every 4 years since 1995. Currently more than 60 countries are participating in TIMSS. Korea has been participating in TIMSS since the first cycle of 1995. However only 8th graders in Korea have continuously participated in TIMSS from 1995 to 2007, and 4th graders participated only in 1995 and not in 1999, 2003, or 2007. They resumed participation in TIMSS in 2011. The following is the summary of the TIMSS results for Korea (Martin, Mullis, Foy, and Stanco 2012; Kim et al. 2012a, b).

The science achievement scores of the Korean 4th Grade students were 576 in 1995 and 587 in 2011 (Fig. 6.1). Those scores were the highest among the participating countries. The average of the TIMSS was set at 500.

TIMSS has four achievement levels; Advanced International Benchmark (625), High International Benchmark (550), Intermediate International Benchmark (475), and Low International Benchmark (400). The countries with the highest average science achievement also were the countries with the largest percentages of students reaching the Advanced International Benchmark. Twenty-nine percent of the Korean students reached this advanced level of performance and 73% reached the High

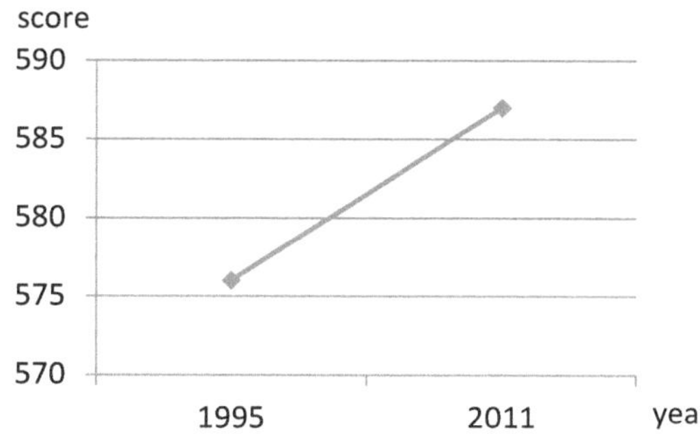

Fig. 6.1 Trend in science achievement of Korean 4th Graders on TIMMS

International Benchmark. Only 1% fell below the Low International Benchmark in TIMSS 2011. Singapore, another top-performing country had 33% of its students reach the TIMSS Advanced International Benchmark. Finland and Japan, other top performing countries had 20% and 14% of their students reached this level, respectively. The percentage of the students reaching the high achievement benchmarks has increased since TIMSS 1995. In TIMSS 1995, 22% of students reached the advanced benchmark and 67% reached the high international benchmark.

TIMSS consists of content and cognitive domains. The content domain consists of Life Science, Physical Science, and Earth Science and the cognitive domain encompasses knowing, applying, and reasoning skills. Knowing covers the student's knowledge of science facts, procedures, and concepts while applying focuses on the student's ability to apply knowledge and conceptual understanding in a science problem situation. Reasoning goes beyond the solution of routine science problems to encompass unfamiliar situations, complex contexts, and multi-step problems. Korean students performed better in Physical Science (597) and Earth Science (603) than in science overall (587), and less well in Life Science (571) in TIMSS 2011 (Fig. 6.2).

By providing data on science achievement, TIMSS also provides contextual information on achievement in science. According to the results, Korean 4th grade students' confidence in science and attitudes toward science were very low compared to students in other countries. Only 15% of Korean students expressed confidence in their science ability, but 43% of international students did so. In addition, 40% of the Korean students did not feel confident in their science learning in contrast to 21% of students in other countries. The gaps between confidence in Korean students and the international average are quite big. At the same time, students' attitudes toward science was determined by their responses to statements such as "I enjoy learning science", "Science is boring", and "I learn many interesting things in science". The international average percentage of students who liked learning science was 53%; only 39% of Korean students agreed that they liked learning science. The percentage of students internationally who do not like learning science was 12% while the percentage of those Korean students was 16%. It is noticeable that some of the highest-performing countries have the smallest percentages of students who liked learning science, such as Korea, Chinese Taipei and Japan.

TIMSS also investigated how confident teachers were in their science teaching and how they taught. Teachers were asked how confident they felt when answering

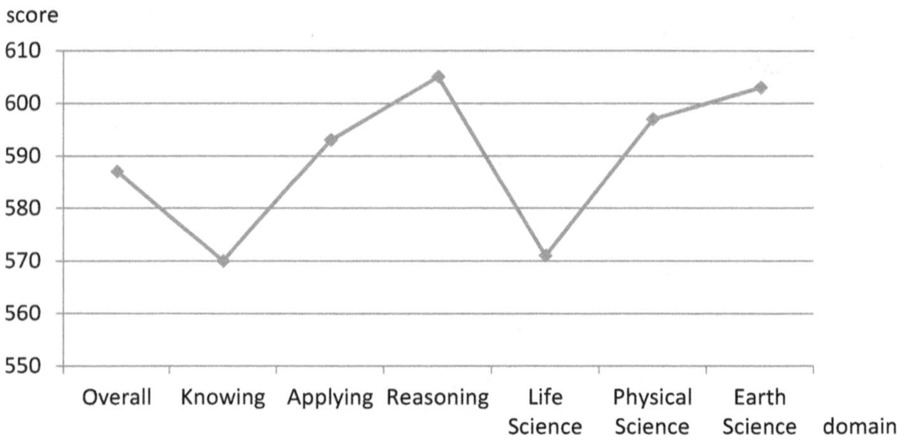

Fig. 6.2 Korean 4th grade students achievement in science by domain

students' questions about science, explaining science principles or concepts by doing science experiments, assigning challenging tasks to capable students, adapting their own teaching to engage students, and increasing students' appreciation of learning science. Internationally, an average of 59% of students had teachers who felt very confident while only 42% of Korean students did so. In general, Korean elementary science teachers are less confident in their ability to teach science than are teachers from other countries.

Regarding their teaching, teachers were surveyed on how often they used teaching practices to engage students in their learning such as summarizing the lesson's learning goals, asking questions to elicit reasons and explanations, encouraging students to show improvement, praising students for good effort, and bringing interesting materials to class. It was reported that 71% of students (the international average) had teachers who tried to engage them in instruction by using a variety of strategies in most lessons. Similarly, 69% of Korean students reported having teachers who used various teaching strategies to engage them in most lessons. Students were also asked how much their science lessons had engaged them. Internationally, on average, 45% of students reported being engaged during their science lessons, another 47% reported being somewhat engaged, and only 8% reported being not engaged. In contrast, only 19% of Korean reported being engaged and 23% were not engaged. The number of students who thought that their science lessons had engaged them is smaller in Korea than other countries though the percentage of Korean teachers who made efforts to engage students in most lessons was similar to that of other countries. Teachers were also surveyed on how often they asked students to conduct science investigations. On average, 40% of international students were taught by teachers who emphasized scientific investigations in half of the lessons or more, and 60% had teachers who had emphasized investigation less frequently. But, 58% of Korean science teachers used instructional practices related to science investigation in about half the lessons or more, which was higher than the international average.

According to the TIMSS results, Korean students showed high achievements in science. Attitudes toward science and confidence in learning science, by contrast, are low (Kim et al. 2012a). Science educators in Korea are understandably concerned about these results and would like to improve attitudes toward science of Korean students. Thus, enhancing curiosity and interest towards science has been made a priority in the goals of the new 2015 national science curriculum (Ministry of Education, Korea 2015). Also according to the TIMSS survey results, Korean elementary teachers made efforts to engage students in science lessons and emphasized science investigation similar to teachers in other countries, but were less confident in their own teaching. Thus, there needs to be more attention and reflection on how to develop classroom practices based on teacher knowledge and self-efficacy throughout elementary teacher education programs here.

6.4 The Development of Elementary Science Curriculum: Historical Background

The government in Korea has centralized the development and revision of curriculum since the first curriculum was issued in 1954. Since then, there have been 10 curriculum revisions and the most recent revision was completed in September 2015. After a new version of the national curriculum has been developed and announced by the Ministry of Education, educational offices at the municipal and provincial levels develop guidelines for curriculum organization and management to reflect local situations and uniqueness. Based on this guide, schools develop school-level curricula to address the contexts and demands from their local communities. Yet, the school-level curriculum is not much different from the centralized national curriculum since the culture of the local community and parental demands are not drastically different from school to school and the influence of and interdependence toward the national curriculum still remain strong in Korean education systems. Furthermore, elementary science textbooks are developed based on the national curriculum as government –designated books; thus, it is convenient for schools and teachers to follow the national curriculum as the main framework and only later include the uniqueness of the local community wherever necessary or suitable. Due to all these reasons, there are not many differences between the national curriculum and school-level curricula in schools.

The Korean national curriculum has undergone a few key stages of development and revision.

The initial stage can be called the stage of instructional syllabus. This stage (1946–1954) was the first decade after the Korean independence from Japan in 1945. At this stage, the curriculum consisted of only content themes and titles of units for each subject area. The first stage of the curriculum was from 1954 to1963. The curriculum at this stage was influenced and developed by the perspective of progressivism; thus, it emphasized everyday life experiences. The second stage of the curriculum (1963–1973) emphasized the instrumental and pragmatic approaches of education more than the earlier stage. The third stage of curriculum (1973–1981) was much influenced by the educational reforms in the US after the Sputnik shock. Based on the criticism of everyday experience-based curriculum, the curriculum at this stage emphasized "going back to basics" by providing structured science knowledge and instructional strategies of science inquiry. However, this resulted in heavy load of science content knowledge in the curriculum, overlooking the development of students. This correction was incorporated in the curriculum reformation at the fourth stage (1981–1987) where the curriculum emphasized human-interests and the holistic development of students as human beings. The fifth stage of the curriculum (1987–1992) incorporated the emphasis of STS (Science-Technology-Society) and continued to stress inquiry-based teaching. The sixth stage of the curriculum (1992–1997) continued the focus on science inquiry and STS education.

In the seventh stage of reform (1997–2007), a few critical changes were made in the curriculum. First, a nation-wide basic curriculum was developed from Grade 1 to high school Grade 1 (Grade 10) and emphasized consistency and relationships of concepts and skills throughout the grade levels. In this curriculum, providing the appropriate curriculum for students' academic levels and needs was emphasized, thus, a differentiated extended curriculum was adapted. During curriculum reforms in 2007, student-centered open inquiry was adopted to emphasize scientific thinking and creative problem solving skills. In the reforms in 2009, the national common basic curriculum was shortened from Grade 1–10 to Grade 1–9 and adopted a new format of "grade groups" and "subject groups." In grade groups, mixed groups of Grades 1–2, Grades 3–4, and Grades 5–6 were introduced, which were managed as separate levels in the previous curricula. Subject groups were categorized based on educational goals and nature of academic disciplines, for instance, social studies/moral education, arts (music/fine arts), and science/home economics were subject groups. In this regard, similarities between science and home economics were recognized. The introduction of subject groups was to facilitate decision-making on allocating teaching hours, to overcome the limitations of separation among similar subject areas, and thus to develop an integrated approach towards knowledge development. Grade groups were suggested to enhance interactions and collaboration between grade levels and to build effective curriculum management in schools.

The most recent curriculum was released in September 2015. The 2015 national curriculum was developed to help students "grow to become creative interdisciplinary human talents with ethical/responsible imagination and scientific technological creativity'. In elementary science curriculum, there are four areas; Motion and Energy, Matters, Life, and Earth and Space and "the travel of water[1]" and "Energy and everyday lives" were created and added as an integrated science approach in elementary science curriculum.

[1] "The travel of water" teaches the water cycle. Curriculum developers named the unit as such to make the unit friendly and engaging for children to learn the concepts here. The unit includes the ideas of how water molecules change their states (solid, liquid, and gas) and also move around in the environment and come back to the earth.

6.5 The 2015 Elementary & Science Curriculum in Korea (as of 2016)

Table 6.1 shows the structure of subject and grade groups and mandatory hours for each subject throughout the elementary years. The 2015 elementary curriculum is divided into two big categories; subject group and optional activities (see Table 6.1). The Subject group in Grades 3–6 includes Korean language, Social studies, Moral education, Mathematics, Science, Practical Arts (Technology, Home Economics), Physical education, Music, Fine Arts, and English. In Grades 1–2, there are five subject groups with more integrated approaches for teaching, that is Korean Language, Mathematics, Disciplined Life, Intelligent Life, and Pleasant Life (see Table 6.1 for more information). Creative experiential learning activities for Grade 3–6 students include learner autonomy, club activities, volunteer work, and vocational activities. For Grade 1–2 students, Everyday Safety is emphasized instead in the category of Creative experiential learning activities.

Science is taught from Grades 3–6 with three lessons per week lasting 40 min each (120 min per week) whereas science in Grades 1–2 is embedded within the subject called Intelligent Life, which is an integration of science and social studies.

Table 6.1 The structure of subject groups and numbers of lessons in a school year

Groups		Grade 1-2	Grade3-4	Grade 5-6
Subject groups	Korean Language	Korean Language 448	408	408
	Social studies / Moral education		272	272
	Mathematics	Mathematics 256	272	272
	Science/ Home Economics	Disciplined Life 128	204	340
	Physical education	Intelligent Life 192	204	204
	Music/Fine Arts		272	272
	English	Pleasant Life 304	136	204
	Sub-total of lessons	1,408	1,768	1,972
Creative experiential learning activities		336 Everyday Safety (64)	204	204
Total number of lessons		**1,744**	**1,972**	**2,176**

1. The duration of one lesson is 40 minutes but is flexibly managed according to seasons, weathers, students' academic levels, the depth of lesson learning objectives, and school situations.
2. The numbers of lessons are for the two-year period of grade group (i.e., Grade 3-4).
3. The total numbers are the minimum numbers of lessons.
4. The lesson numbers of Home Economics is only included in Grade 5-6 science/home science group

This subject focuses on inquiry skills such as observation, classification, investigation, prediction, and drawing on relationships among things or ideas and the eight core themes of school, family, local community, nation, Spring, Summer, Fall, and Winter. Each theme includes detailed related subthemes, for instance, for Spring there are sub-themes of the weather and lifestyles in Spring, everyday tools in Spring, Spring gardens, and Plant growth. The eight core themes are the same for Disciplined Life and Pleasant Life.

The 2015 national curriculum ultimately aims to develop creative interdisciplinary human talents; the latter means that people develop ethical/responsible imagination and scientific and technological creativity together with right character, develop new knowledge and create new values by integrating knowledge (MOE 2015). These terms representing the goals are stronger and more evocative in the original Korean language than in English, which is a pity as they are hard to translate. Nonetheless, to achieve these goals, the curriculum describes the details of desirable human moral characters and core competencies of creative interdisciplinary human talents. The science curriculum is aligned with the overall elementary curriculum by stating core competences in science. Scientific thinking, scientific inquiry skills, scientific problems solving skills, scientific communication skills, and scientific participation and life-long learning skills are thus emphasized as subject core competencies in order to address the goals of enhancing creative interdisciplinary human talents.

Whereas science is taught in an integrated subject in Grades 1–2, it becomes a standalone subject from Grades 3–6. Thus, the details and structures of science units and learning objectives are provided only from Grade 3 in elementary science curriculum. The units of science curriculum in Grades 3–4 and Grades 5–6 are as follows. In each grade group, there is one unit of "integrated science"; the "travel of water" (Grades 3–4) and "Energy and everyday lives" (Grades 5–6). These two units are new additions to the previous science curriculum. Each unit in the curriculum contains the overall goals and directions of the unit, achievement standards (i.e. learning objectives), inquiry activities, instructional strategies and guidance, assessment strategies and guidance. Compared to the previous curriculum, instructional and assessment strategies and guidance for teaching are new additions in the 2015 curriculum (Table 6.2).

Inquiry activities are suggested to help teachers and students achieve achievement standards more effectively. The following example shows how inquiry activities are introduced in the unit of shadows and mirrors in Grades 3–4 (Table 6.3).

By including a separate section on inquiry activities, the emphasis of inquiry becomes evident in the curriculum structure and classroom practice. Yet, this structure has also been criticized for the lack of recognition concerning attitudes as important learning objectives such as interest and curiosity toward science, scientific attitudes, and the joy of learning science (Na, Yoon, and Kim 2015). Given that the levels of students' positive attitudes toward science and science learning have been reported as low in international scientific literacy test such as Programme for International Student Assessment (PISA) (Kjærnsli and Lie 2011; Lee, Sohn, and No 2008) and TIMMS (Martin, Mullis, Foy, and Stanco 2012), this critique needs to be thoughtfully taken into account for further development. In the present curriculum, students' attitudes toward science are mentioned in the overall goals of science curriculum, not in the students' achievement standards of each unit.

Table 6.2 The units in the 2015 elementary science curriculum in Korea

Grades 3–4	Grades 5–6
1. Characteristics of materials & substances	1. Temperature and heat
2. Usage of magnets	2. The solar system and stars
3. The lives of animals	3. Dissolution and solution
4. Changes of the earth surface	4. Various living beings and our lives
5. Lives of plants	5. Life and the environment
6. Strata and fossils	6. Weather and our lives
7. States of substance	7. Movements of objects
8. Characteristics of sound	8. Acids and bases
9. Weight	9. Movements of the earth and moon
10. Lifecycles of animals	10. Various gases
11. Volcanoes and earthquakes	11. Light and lenses
12. Separation of mixtures	12. Structures and functions of plants
13. Lifecycles of plants	13. Usage of electricity
14. Changes of states of water	14. Seasonal changes
15. Shadows and mirrors	15. Combustion and fire extinction
16. Shape of the earth	16. Structures and functions of our bodies
17. The travel of water (integrated topic)	17. Energy and living (integrated topic)

Table 6.3 The achievement standards and inquiry activities in the unit of shadows and mirrors

<Achievement Standards>
[4 sci 15–01] Explain how shadows are formed by observing the shadows of various objects.
[4 sci 15–02] Observe and describe the changes of size of shadow according to the distance between an object and light.
[4 sci 15–03] By comparing object and its reflection on mirrors, explain the characteristics of mirrors.
[4 sci 15–04] Investigate the examples of using mirrors in everyday lives and explain the function of mirrors relating the characteristics of mirrors.
<Inquiry activities>
Comparing shadows of transparent objects and opaque objects.
Observing the changes of shadow size according to the distance between a light source and object.
Comparing an object and the reflection of it on a flat mirror.

6.6 Elementary Science Textbooks

Educational books in elementary science include science textbooks, activity or work books, and teacher guidebooks. According to the regulations on educational books, there are three kinds of educational books; Government-designated books, authorized books, and approved books. Government-designated books means that the Ministry of Education has developed and has the copyright of it. When a book is developed by the third party and later authorized by the ministry, it is a government authorized book. When a book is not verified by the government, but still acknowledged for use in schools, it is called as government approved book. In elementary schools, textbooks of Grades 1–2 Korean Language, Mathematics, Disciplined Life, Intelligent Life, and Pleasant Life and textbooks of Grade 3–4 and 5–6 Korean Language, Social studies, Moral education, Mathematics, and Science are government-designated books and the rest of the textbooks such as Music or Physical Education books are government authorized books.

Science textbooks for the 2007 and 2009 national science curriculum were developed by the Korea Foundation for the Advancement of Science & Creativity (KOFAC), which was designated to do the job by the Ministry of Education. The set of main science textbooks, science activity books, and teacher guidebooks were developed and distributed nationwide. This set is the only kind of elementary science textbook, thus, teachers' dependence on this is extremely high. Science activity books provide processes of practical work, worksheets, and questions for further development aligned with textbook topics and activities. They often function as science notebooks. Teachers and students use the activity books as learning and assessment tools. The process of developing science textbooks is explained below (see Fig. 6.3). During the development of textbooks, the MOE and designated team of textbook development examine the quality and effectiveness of various drafts until they produce the last draft for field test. The last draft is distributed to several schools for testing the quality and practicality of the book and is revised for the final version. Sixteen elementary schools participated in the testing process for 2009 science curriculum-based textbook development. Then the final versions are distributed nationwide. Science textbooks (main science textbooks, science activity books, and teacher guidebooks) for 2015 national science curriculum will be developed through a similar process.

The process of textbook development explained above shares both strengths and limitations. The process is unified and controlled by the government, thus, it could provide the same levels of content and quality of the textbooks across the nation. Accordingly, this could reduce the gap of educational experiences among students in various regions. For instance, in a small remote community, teachers could encounter a lack of teaching resources. Some teachers might not be interested in teaching science, thus, not striving to develop teaching materials. For these cases, the distributed textbooks across the nation would help teachers with teaching ideas and resources; thus, reduce the gaps of educational quality and achievement among different regions with challenging situations.

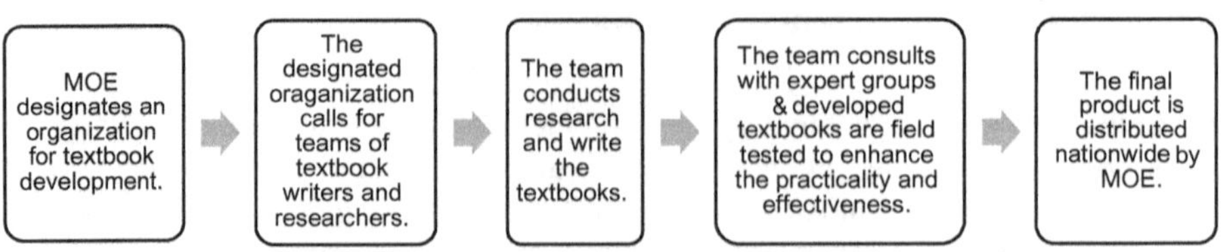

Fig. 6.3 The process of science textbook development

However, there are also concerns about the unified national science textbooks. That is, the same content of science textbooks do not encourage teachers' flexibility and creativity on curriculum decision-making. All children in Korea learn the same content, same activities, and same ideas and this could jeopardize the level of creative thinking and problem-solving skills for children and the nation. Also, teachers depend on the textbooks too much, not understanding the goals of science curriculum. They simply regard textbooks as the curriculum although textbooks are certainly not the curriculum! They are part of instruction as a means of achieving the goals of the curriculum. Thus, teachers would need to understand how to use textbooks more effectively and creatively, not depending heavily on the textbooks only. The degree of importance of textbooks and teachers' dependence on them varies. In the Korean situation, the degree of dependence and importance is high since textbooks are the government-published book and there is only one kind of textbook for elementary science across the nation, which brings out the notion that textbooks are treated as a practical version of the curriculum among teachers (Sim et al. 2010)

6.7 Teacher Education Programs in Higher Education

The teaching profession is one of the most popular career choices in Korea. Traditionally, teachers are well respected in East-Asian cultures including Korea (Liu et al. 2015) and they are government employees with secure job status and competitive salary. When s/he is hired as teacher at a national or public school, s/he as a civil servant will work till they retire with a pension. Thus, under the current climate of job markets and unemployment rates, teachers became one of the most desirable jobs among young people, especially female students. The level of desirability toward teaching profession has led to intense competition among high school graduates to enter teacher education programs, which is determined by the scores of the national university entrance exam. The Ministry of Education also determines the quota of the first-year students to elementary teacher education programs each year based on the demographic changes of children, teachers, and schools. Despite government attempts to control the supply and demand of teaching positions, teacher oversupply already occurred at elementary schools a decade ago due to the unpredictable numbers of student population and below replacement fertility rates in Korea. The average employment rate of graduates from elementary teacher education institutions was 55.1% in 2011, 64.7% in 2012, and 66.5% in 2013.

Teacher education programs for elementary education in Korea are provided in designated universities that focus only on elementary teacher education and teacher certificate programs. There are 13 designated institutions across the nation; 12 national universities (10 national universities of Education and two national universities), and one private university. It is a unique situation in teacher education programs in Korea that such universities only specialize in elementary teacher education and all students in the universities aim to become elementary school teachers or work in the areas of elementary education after graduation. This system could be effective to develop specialization and professional knowledge and skills in elementary education; yet, such independence and separation could also raise some concerns on lack of opportunities to change career tracks for students.

Once students (i.e. teacher candidates) are admitted to teacher education programs in such universities, they complete the four-year program to obtain their teaching certificates. Students enter the program as one cohort and then they select their major courses among 12–14 subject areas: Korean, Ethics, Mathematics, English, Social Studies, Science, Music, Arts, Physical Education, etc. Even though they choose one specialization out of these, they learn all other subject areas to competently teach all subjects in elementary schools. After the four-year teacher education program, students obtain their teaching certificates, but this does not guarantee employment in schools. Teaching certificate holders must now pass the National Teacher Employment Test in order to teach in public schools. The National Teacher Employment Test is administered annually by the Korea Institute for Curriculum and Evaluation, a government-funded educational research institution and each Metropolitan and Provincial Office of Education. Upon passing this test, teachers gain tenure to teach in public elementary schools, thus, this test is extremely competitive and challenging. The structure of the Test is described next.

The test consists of two parts; (1) written tests of essay writing on educational theory and practice (20%), 22 questions related to curriculum issues (80%), and Korean history (this is a Pass/Fail test separately organized and verified by the government), and (2) an in-depth interview on teaching aptitude, teaching practice, English language abilities, and English teaching. In the first test, there are 22 curriculum-related questions (three questions in Korean language, three in mathematics, two in social studies, science, English, and physical education, and one question each in moral education, practical arts, music, fine art, and other general areas of elementary education). In science, teacher candidates are given an example of a lesson plan and a brief description of lesson progress with classroom dialogues between students and teacher and teacher candidates need to interpret students' learning and students' misconceptions, anticipated students' inquiry skills, etc. from the given description and information. For this kind of questions, teacher candidates should know specific scientific concepts, misconceptions, and types of inquiry skills and should know how to interpret classroom situations and dialogues. Since the test is highly competitive and critical for students' future and career development, the validity of test design and items are very sensitive and debatable (Kang and Ahn 2014; Lee et al. 2013). Thus, the government puts much effort to secure the validity and authenticity of the test each year. Teaching certificates can be obtained upon the completion of professional courses in 4-year teacher education programs according to the Teacher Certification Authorization Act (Act No. 25684) and its detailed regulations. Therefore, the curricula structure of teacher education and teaching certificate programs are mostly controlled and influenced by the government in Korea.

Table 6.4 shows the structure of the curriculum of three typical national universities of education for elementary teacher candidates. A minimum of 140 credits are required for graduation; there are about 20 credits in general pedagogy, about 50 credits in content specific pedagogy, and about 20 credits in major courses of special-

Table 6.4 Credit hours of curricular domains of universities of education (as of year 2016)

Curricular domains	Chuncheon national univ. of education	Gyeongin national univ. of education	Seoul national univ. of education
Liberal arts course	34	36	40
Pedagogy	19	21	18
Content specific pedagogy	49	47	50
Creative experiential activity	2	2	
Subject practice	12	12	8
Intensive course	21	21	20
Practicum	4 (10 weeks)	5 (10 weeks)	4 (9 weeks)
Graduation thesis or graduation task	P/F	P/F	P/F
Educational volunteer program	P/F	P/F	P/F
Aptitude and personality of teaching test	P/F	P/F	P/F
Total	**141**	**144**	**140**

ization. National universities of education have 12–14 major courses, of which the Elementary School Science Education department is one. Therefore, if students are from the Elementary School Science Education department, they will get 20 credits about elementary science education including Elementary School Science Inquiry, Elementary School Science Teaching Resources, etc. On the other hand, students from different departments, for example Elementary School Music Education department, will get only about 5–6 credits of science education among 50 credits concerning content specific pedagogy before graduating. The teacher education curriculum in the most of the universities includes two to three times of school practicum during 9–10-weeks in total. For example, Chuncheon National University of Education (CNUE) requires one-week observation in the spring and fall term of the second year.

During the practicum, students need to spend some time at a special education school. In the fall term of the third year, a four-week practicum is offered and students prepare and teach lessons and work with their mentor teachers. They do a comprehensive practicum in the spring term of the fourth year which they practice everyday school teaching such as classroom management and government document management. Usually every student teacher teaches 15–20 class hours during 4 weeks of the practicum; thus, they may teach one to three science lessons for the entire practicum. This is not enough to develop pedagogical strategies and confidence in science teaching, especially pre-service teachers showed great difficulties in conducting science practical work in classroom teaching and found the gap between university instructions and actual teaching with science experiments (Yoon 2004). This challenge is also closely relevant to the issue of elementary teachers as generalist versus specialist. Like other countries, elementary school teachers in Korea usually teach most of the subjects; thus, they go through generalist tracks.

This means there is not enough time to develop science knowledge and to practice science teaching during practicum. The debate on specialist and generalist in elementary education is still on-going.

Another issue in teacher education program worth discussing in this section is that elementary and secondary teacher education programs are offered in different institutions in Korea (Im, Yoon, and Cha 2016). In most countries, elementary and secondary teacher education programs are offered in the same faculty of education, but in Korea universities for elementary teacher education are independent and separated from other universities with secondary teacher education programs. Having independent specialized institutions for elementary teacher education has pros and cons; this could develop the professionalization of elementary curriculum and pedagogy, effectiveness of program focus and material circulation, maximization of system management, etc. but there have been also some concerns in the current years. The isolated institutions from other universities could result in some degree of isolation among educational programs and lack of collaboration with secondary school educators and researchers.

There has been also an emerging concern around elementary teacher education program, that is, challenges and pressure on rethinking the track of becoming an elementary teacher. Since elementary teacher education programs and teaching certificates are specialized in specific universities, it is not easy to become elementary teachers without completing the programs in such institutions. But recently, there have been discussion and external pressure on making elementary teacher education programs more open and accessible to those with specific content knowledge and skills through various educational institutions who want to become elementary school teachers. For example, people with Bachelor degrees in the sciences have asked for policy and institutional reformations to become science teachers in elementary schools. Yet, this also raises concerns about teacher un/employment rates and the lack of pedagogical professional knowledge in classrooms, thus thorough examination and approach is needed.

6.8 Assessment in Elementary Science Education

Schools and teachers have autonomy in decisions related to their students' assessment. The purpose of assessment in elementary schools should be to improve students' learning and the quality of instruction. In order to achieve the purpose of assessment, the Korean national curriculum has provided direction and guidelines for assessment practices. In the new 2015 national curriculum, the latter were strengthened. Whereas in the past national curriculum, the instructions for assessment were provided at the end of the curriculum, in the new national curriculum, assessment methods and strategies were provided for each unit. These assessment methods and strategies include suggestions for conducting quality assessment such as appropriate assessment methods for the unit, what should be assessed in the unit, and possible pitfalls of assessment. One example of the assessment methods and strategies for the unit, "The Properties of Magnets" in the national science curriculum for Grades 3 and 4 is as follows:

> Performance assessment. Observing how students are performing the tasks can be implemented while students are doing their inquiry activities.

The new national curriculum involves the directions and notes of assessment at the end to provide general principles and strategies for assessment as follows.

> In 'Science, the understanding of basic concepts, science competencies through inquiry activities, and scientific attitudes are assessed in balance, and thereby, the following are stressed:
>
> 1. Assess the understanding of basic concepts and the ability to apply them.
> 2. Assess science competencies including scientific thinking skills, inquiry abilities, scientific problem solving skills, science communication skills, engagement in science, lifelong learning skills.
> 3. Assess the interest in science and the recognition of the values of science, active participation in science learning, collaboration, the attitude to solve problems scientifically, and creativity, etc.
> 4. Assess with the use of various methods including multiple-choice tests, essay type and writing tests, observation checklists, reports, performance tests, interviews, and portfolios, etc.
> 5. Use topics and situations that can improve creative problem-solving skills and character cultivation.
> 6. Assess group performance as well as individual performance to cultivate cooperation.
> 7. Develop assessment instruments in collaboration between teachers for better reliability and validity.
> 8. Assessment is consistent with achievement standards, and assessment data is used for planning lessons, improving teaching strategies and guiding students' careers.
> 9. Assessment is conducted following the steps of making plan, developing tests and instruments, implementing assessments, processing assessment data, utilizing results, etc. (Ministry of Education, Korea 2015).

The new national curriculum assesses not only the understanding of basic concepts but also science competencies. Science competencies are newly introduced in the most recent national curriculum. Science competencies are key abilities and skills that students should have for successful performance in their learning and everyday lives; scientific thinking skills, inquiry abilities, scientific problem solving skills, science communication skills, engagement in science, and lifelong learning skills. Science competencies are recognized as learning outcomes; thus, are to be assessed in science classrooms.

There is another document provided at the national level to assist teachers to improve teaching and assessment. The Achievement Standards and Achievement Level for each subject are provided at the national level in tandem with the national curriculum. The achievement standards and level are to help teachers conduct quality teaching and assessment. The achievement standards define what students should know (what teachers should teach) and be able to do (what teachers aim for in their teaching). The achievement level describes how well students met the achievement standards at three levels—basic, proficient, and advanced levels. Elementary school teachers occasionally adapt these to four levels—need improvement, expected, good, and excellent. Elementary school teachers are expected to utilize the achievement standards and levels based on the national curriculum when they assess students' learning and development in their classrooms.

Elementary schools normally conduct their own academic achievement assessment based on the policy of metropolitan/provincial offices of education. Municipal and provincial offices take decisions on important educational matters including assessment policy. There has recently been a tendency to abolish assessment in the form of mid-term or final examinations. Several municipal and provincial offices have abolished mid-term and final examinations and encouraged formative assessments in the form of performance assessment. Decisions on assessment in elementary schools are made by committees for student record management in each school. Though there may be differences among schools, it is common that schools plans and establish yearly assessment plans along with their overall yearly educational planning. Yearly assessment plans include details of the regular testing period for each subject, frequency and methods of assessment, the extent to which assessment results are reflected, and the test grading methodology and results utilization plan. Assessment results are provided in the form of "Student School Records". The student school records provide detailed information about student academic performance by subject.

There is one nationwide assessment, the "Subject Learning Diagnostic Test (SLDT)" for elementary students in Korea. The SLDT is provided by municipal and provincial offices of education and administered to 4th-, 5th- and 6th-graders. The purpose of SLDT is to analyze student achievement, identify those who have fallen behind, and provide responsible instruction to help them acquire basic academic competence. The SLDT includes five subject areas: Korean language, social studies, mathematics, science, and English. Assessment questions take the form of multiple choice and short answer questions. Student achievement levels are identified as "achieved" or "not achieved" in assessment reports but the results of this test are mainly used for helping struggling students and setting up remediation plans. In addition, assessment results are reflected in establishing education policies to reduce the number of students who have fallen behind and improve the quality of teaching and learning.

Until 2012, the National Assessment of Educational Achievement (NAEA) was utilized annually for Grade 6 in primary school in Korea. From 2013 NAEA is conducted only for Grade 3 in middle school and Grade 2 in high school. The main objectives of NAEA were to measure educational achievement and to analyze the trends of student achievement.

However, as the objection and resistance toward the national standardized tests were growing the Korean government decided not to continue NAEA in elementary schools. The opponents of NAEA claimed that NAEA orients teachers to teach mainly for the test. They pointed out that NAEA interfered with balanced teaching, and created stress for students and teachers. In addition, they insisted that NAEA resulted in excessive competition between regions and schools and led to expansion of private education when the results of the assessment were released (Korea Institute of Curriculum & Evaluation 2012).

In Korea, policies regarding national assessment have changed several times as criticism of standardized testing has grown. At first, the national assessment for primary students started for only sampled students to obtain the information of student achievement and educational programs. After several years, the national assessment for elementary students became tests for the whole population. Criticism of the national tests increased and finally the national assessment for elementary students was abolished in 2013. Opponents of standardized tests insist that the tests led to negative outcomes such as content-centered teaching, yet, there are still those who insist the necessity of national assessment to collect information for improving students learning and educational systems in the nation. In order to be meaningful, assessment should be carefully designed and implemented. Teachers' expertise in assessment is necessary for accurate and quality assessment. Many primary school teachers find it difficult to design and implement assessment, especially performance assessment. As assessment is essential to enhance not only student learning but also the quality of teaching and educational programs, the Korean government has been providing support to improve teachers' expertise in assessment.

6.9 Teacher Practice of Assessment in Science Classrooms

There are no mandatory methods or structures of assessment in elementary science classrooms in Korea; thus, school principals and teachers decide frequencies, methods, and types of assessment based on the situations of school and community and parents' demands. There is a great level of flexibility in teachers' decision-making on classroom assessment, yet, there are also common goals and practices of assessment across schools and classrooms. That is, teachers aim to evaluate students' science knowledge and inquiry competencies through various types of assessment such as pencil and paper tests, practical tests, science writing, lab reports, etc. Written tests such as unit tests, mid-terms, and final term tests are commonly practiced in science classrooms. Teachers do not give a unit test for each unit unless there is concern about students' understanding of certain concepts and skills in the unit. Mid-term and final term tests are more commonly practiced in classrooms. Yet, the actual scores or test results are used for teaching and learning development, not recorded on report cards for students and parents. Recently the tendency of reducing mid-term or final examinations emerged, thus, it is expected to see less practice of these types of exams in elementary classrooms and more of formative, performance-based assessment to understand and enhance students' learning.

To increase alignment among curriculum, instruction, and assessment, the core achievement standards are distributed to schools and teachers are encouraged to utilize these standards in their teaching and evaluation. The core standards of achievement are developed based on the learning objectives in the curriculum to develop the alignment of curriculum and assessment. Teachers decide the details of assessment plans and types based on the standards and the levels of achievement. The table blow shows how teachers plan classroom assessment for the first term in Grade 5 (Table 6.5).

Teachers share various teaching resources such as lesson plans, teaching materials, and test items in several educational websites and blogs. With the development

Table 6.5 An example of assessment plan on achievement standards (Grade 5, Term 1)

Area	Unit	Achievement standards	Levels	Indicators for achievement levels	Types of assessment	Timeline
Earth & Space	1. Earth and the moon	Students understand the shapes and characteristics of the earth and moon and know the reasons why there are day and night and why the moon changes its location and shapes.	Excellent	Students understand the shapes and characteristics of the earth and moon and correctly explain the reasons why there are day and night and why the moon changes its location and shapes.	Students' descriptive reports	2nd week, April
			Good	Students understand the shapes and characteristics of the earth and moon and briefly explain the reasons why there are day and night and why the moon changes its location and shapes.		
			Expected	Students understand the shapes and characteristics of the earth and moon and explain the reasons why either there are day and night or why the moon changes its location and shapes.		
			Needs improvement	Students understand the shapes and characteristics of the earth and moon.		
Life	3. Structures and functions of plants	Students understand and explain the structures and functions of roots, stems, leaves, flowers, and fruits and how they are interrelated.	Excellent	Students understand and correctly explain the structures and functions of roots, stems, leaves, flowers, and fruits and how they are interrelated.	Students' verbal presentation	3rd week, June
			Good	Students understand and briefly explain the structures and functions of roots, stems, leaves, flowers, and fruits and how they are interrelated.	Teachers' observation on students' activities	
			Expected	Students understand and explain the structures and functions of roots, stems, leaves, flowers, and fruits.		
			Needs improvement	Students show interests in the structures and functions of roots, stems, leaves, flowers, fruits but do not understand and explain them.		

Area	Unit	Achievement standards	Levels	Indicators for achievement levels	Types of assessment	Timeline
Matters	2. Dissolution and solution	Students understand the basic concepts of dissolution and solution and know the relationship between quantity of solute and solvent.	Excellent	Students understand the basic concepts of dissolution and solution. Students know the relationship between quantity of solute and solvent to conduct an experiment and draw a conclusion.	Teachers' observation on students' activities	2nd week, July
			Good	Students understand the basic concepts of dissolution and solution and know the relationship between quantity of solute and solvent.		
			Expected	Students understand the basic concepts of dissolution and solution and participate in an experiment.	Written test (pen & pencil test)	
			Needs improvement	Students do not understand the basic concepts of dissolution and solution and minimally participate in an experiment.		
Science inquiry	Inquiry activity	Students work on a group practical work and summarize and present findings.	Excellent	In order to solve inquiry problems, students make tables of results and interpret the meanings of the findings in details and accuracy.	Performance-based assessment: Observation on students' practical work	4th week, April
			Good	In order to solve inquiry problems, students make tables of results and briefly interpret the meanings of the findings.		
			Expected	In order to solve inquiry problems, students make tables of results but do not interpret the meanings of the findings.		
			Needs improvement	In order to solve inquiry problems, students show difficulties in making tables of results and interpreting the meanings of the findings.		

of social network system and computer technologies, this type of resource sharing is becoming more popular among teachers. Many materials and resources shared in teacher resource websites are often developed, modified, and examined by teachers for their own needs and convenience. The organization of "Science All" managed by the Korea Foundation for the Advancement of Science & Creativity (KFASC) also provides a website for teachers to provide useful information and resources for teaching and assessment (http://www.scienceall.com). Teachers often download lesson ideas and test items from the website. The example in Table 6.6 is an assessment resource for Grade 3 science from the KFASC website. The resource on the website is in Korean and the authors translated it into English.

This example of test item analysis is useful for teachers to understand what each test item focuses on in terms of students' knowledge, inquiry skills, and attitudes, creativity, the levels of difficulties, and types of assessment. When teachers use these resources, they often change the formats and contexts of test items for their own needs and classroom contexts. When they used the item above, they changed the sun-looking light source to a light bulb to increase the options of answers. They thought students would think the light was the Sun, then, they could not remove the Sun or change the location of the Sun, thus, this would limit the answer of removing the light source.

Teachers also use various resources for performance-based assessment. They use observation and checklists while students participate in practical work. During practical work, students are engaged in various types of science inquiry skills, collaborative decision making, communication, and presentation, thus, teachers attempt to look into students' classroom actions and behaviors. However, performance-based assessment during students' experiments is hard for teachers to evaluate students' abilities accurately. To help with the difficulties of performance-based assessment, teachers often use worksheets or science activity books to assess students' learning and inquiry skills through practical work. Teachers also use various teacher resource websites which are accessible and free for local teachers to share their teaching resources such as lesson plans, activities, and assessment strategies. Figure 6.4 is an example of performance-based test that teachers use in Grade 4 science classrooms. The teacher referred to the various materials from the teacher resource websites and developed the worksheet in Fig. 6.4. Her students learned the relationship between weight, distance of loading points, and balance by manipulating balance bars, weights, clips, and scales. Students were given the worksheet with questions to work on and teachers used this worksheet for practical assessment.

Figure 6.5 shows an example of students' work in a science activity book, which is part of the official science textbooks in Korea. This activity is based on acids and bases in Grade 5 primary science. Based on the curriculum and textbooks, students performed an experiment to classify liquids according to whether they are acids or bases. They were given two pH indicators; litmus paper and phenolphthalein and various test liquids such as vinegar, lemon juice, detergent, soda, etc. They put the litmus paper and a few drops of phenolphthalein into the liquid and recorded results and classified those liquids in the tables in the activity book. The learning objectives of this activity included to observe changes and classify liquids based on color

Table 6.6 An example of planning and analyzing science tests

Item no.	3-2-4-1			
Unit	Light and shadow		Subunit	Understanding light
Topic	Let's think what will happen without light.			
Test focus	Explain how eyes can see objects in relation to light.		Levels	(v) High, ( ) Medium, ( ) Low
Lesson process	( ) Diagnostic (v) Formative		Test type	*Written test*
	( ) Summative			( ) Multiple choice
Knowledge	( ) Know, ( ) Understand, (v) Apply			( ) True false
				( ) Blank
Inquiry skills	*Basic*	*Integrated*		( ) Connection
	( ) Observing	( ) Understanding a problem		( ) Short answer
	( ) Classifying	( ) Hypothesizing		(v) Open ended
	( ) Measuring	( ) Designing & conducting a test		*Performance-based*
	(v) Predicting	( ) Controlling variables		( ) Observation
	( ) Inferring	( ) Interpreting		( ) Interview
		( ) Concluding & generalizing		( ) Report
				( ) Portfolio
				( ) Science writing
Attitudes	( ) Attitudes toward science		Creativity	( ) Uniqueness ( ) Fluency
	( ) Attitudes toward science learning			( ) Elaboration
				( ) Integration
	( ) Scientific attitudes			( ) Simplicity ( ) Relevance
				( ) Consistency

Question item 3-2-4-1

The drawing below shows how our eyes can see the toy with the ways of light. Based on the drawing, explain two ways that we cannot see the toy.

Suggested answers:

Close your eyes. Block the light path. Place the toy behind your back. Remove the light source.

Guiding point:

To see an object, there should be two conditions: light and reflection of light into our eyes.

Thus, either of the conditions should be removed not to see the object.

Source from http://www.scienceall.com

changes. Later, they developed the concept of acid and base. After students finished their experiment and recorded the information, the teacher collected students' activity books to understand students' experiment results and understandings of concepts. The following two pages scanned from the activity book show how science

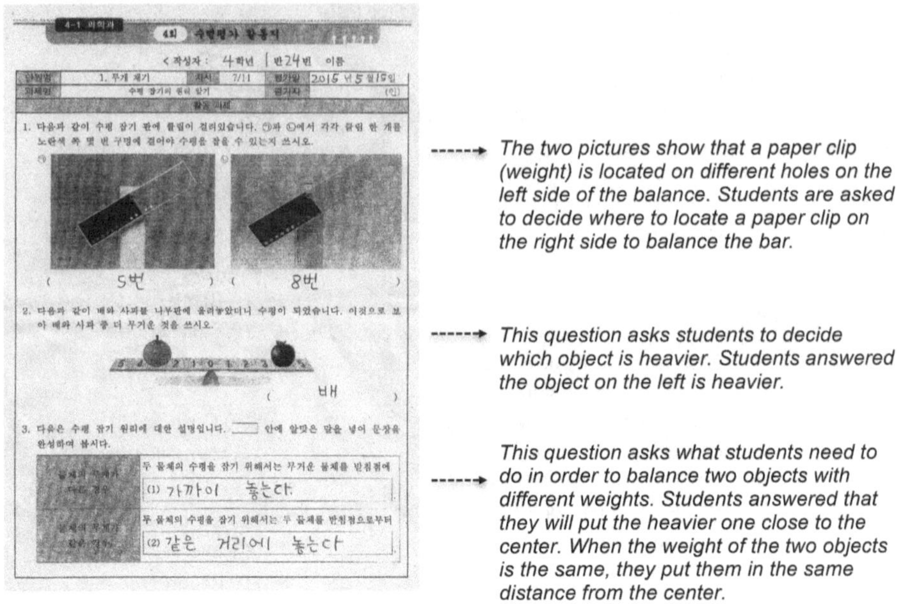

Fig. 6.4 An example of performance-based assessment on practical work

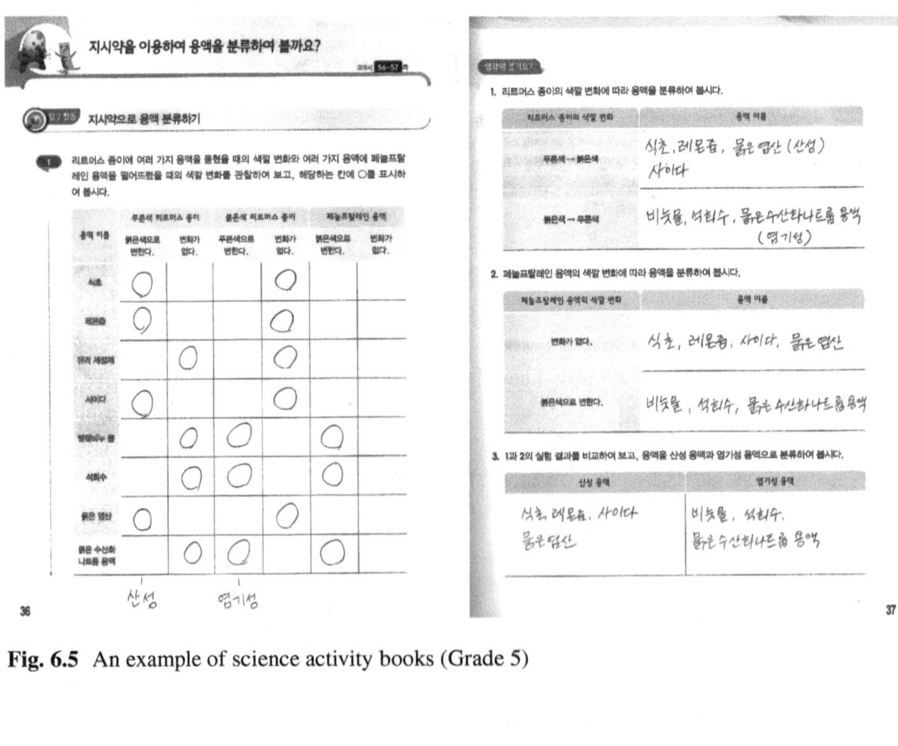

Fig. 6.5 An example of science activity books (Grade 5)

activity books are aligned with classroom activities and learning objectives and how teachers could use them as assessment tools. The page starts with a question, "Shall we classify liquids by using indicators?" and provides tables for students to record their findings.

In the new assessment plan released in 2015, the assessment on science competencies and inquiry skills are more emphasized now in elementary science. There will therefore be more practice on performance-based assessment on students' scientific thinking, inquiry process, creative problem solving, etc. In the 2015 curriculum, core competencies of science are clearly stated, yet, it is not clear how these overall competencies are connected to actual achievement standards of each science unit. There is a need for showing a clear connection between long-term overall goals of science education and the short-term classroom-based achievement standards so that teachers could see how their actions of classroom teaching and assessment are achieving the overall goals of science education.

6.10 Conclusion: Issues and Concerns for Further Development

In this chapter, we have described various ways how elementary science education has been developed over the decades in Korea. There has been much progress in the development of science education, but there are also many challenges to be addressed. As seen in the earlier section, the national curriculum in Korea has been revised frequently, especially since 2000 till now. This has resulted from the continuous interests and concerns of education policy makers and scholars to enhance students' learning and national development in order to prepare for rapid changes in local and global society. This could be a legitimate and positive action for educational development, yet, the frequent revisions and changes result in confusion and frustrations on implementation among school teachers, students, and parents.

Throughout the history of curriculum development here, it is evident that the importance of science inquiry has been recognized and emphasized in students' learning objectives. Since the third national curriculum in the 1970s, science inquiry has been emphasized as key goals of scientific literacy, yet, without a well-developed theoretical foundation of inquiry teaching and feasible pedagogical suggestions in Korean educational and social contexts, its approach has been confusing to many teachers in classrooms. According to research studies, teachers here hold several (mis)understandings of science inquiry, overlook the importance of inquiry teaching (Park et al. 2004; Yoon et al. 2011, 2013) and experience difficulties in inquiry teaching (Jin and Jang 2007; Lee et al. 2007). This calls for further consideration on what would be the appropriate strategies of inquiry teaching to Korean classroom contexts for effective classroom implementation.

Science inquiry and science competences for critical and creative problem-solving are noticeably emphasized in the 2015 national science curriculum. This emphasis requires less learning of science content and instead have more inquiry practices in school. Also, it calls for less paper-and-pencil tests, and more performance-based assessment to develop students' scientific thinking and problem-solving. These changes are positive and desirable although the reduction of content knowledge in science curriculum has not been an easy task among curriculum developers, educators, and researchers in science education. Also, under the current assessment systems, it is challenging for teachers to switch their behavior from content focus to inquiry teaching (Kim, Tan, and Talaue 2013). To encourage changes in teachers' teaching practices, there needs a thorough examination and research on teachers' perceptions on the changes of policies and reflections on changes in their actions (Hall and Hord 2011; Kim and Tan 2011). The curriculum changes need to be critically examined an consulted by/with teachers for further development and effective implementation.

Another issue of elementary science education is an on-going discussion of generalist and specialist science teachers in elementary science classrooms (Kim and Tan 2011). In the current educational system, elementary school teachers teach all subjects and they are trained to be generalists. Students in teacher education programs will choose a specialization in subject areas such as science, math and Korean although they also need to develop pedagogical content knowledge in all elementary subject areas. Opportunities to develop pedagogical skills of science teaching practice during practicum are also lacking. Thus, the focus on specialization in science is not sufficient for them to be specialist science teachers through the teacher education programs.

In order to improve the quality of elementary science teaching, the government has encouraged schools to have specialist teachers for physical education, English, and science. The local offices of education that supervise schools decide how many specialist teachers will be in each school and school principals decide which subjects will be taught by specialist teachers in their schools. However, even the specialist science teachers have some degree of difficulties in teaching science because they were not trained to be specialist science teachers, but rather generalist teachers in their pre-service teacher education programs. Even if they are called specialist science teachers, they are not true specialists in science because they chose to teach science based on interest and willingness. The issue of specialist science teachers is an ongoing discussion and a work in progress in the Korean education system to enhance the quality and self-efficacy of elementary science teaching.

As the TIMSS results have shown, students' interests and motivation in science is another challenge in Korean science education. Students' interests and confidence in science learning are low even in elementary grades and the number of students entering science fields is descending in Korea (Jin and Yoon 2002; Park 2004). As students' positive attitudes are critical aspects of students' scientific literacy and future studies and careers in the field of science and technology, it is critical to cultivate students' curiosity, interests, and enthusiasm toward science learning. As research showed that most students lose their interest in science in elementary grades (Driver et al. 1996), there needs to be certain strategies to address and alleviate this issue tactfully. The new science curriculum 2015 states that the first goal of science education is for "students [to] enhance curiosity and interest in natural phenomena, and develop an attitude to scientifically solve problems in daily life." Without strategic action plans, this goal would be only another add-on statement in the curriculum. In the new science curriculum 2015, inquiry-based and performance-based learning and teaching are encouraged to enhance students' interests and curiosity. And yet, it is uncertain how this encouragement would be taken by teachers and students in the highly competitive educational settings with high stakes exams. The university entrance exam and overheated private education in Korean society have driven teachers', students' and parents' expectation and decision-making on learning outcomes and teaching practice. This conflict is a huge impediment but must be publicly discussed.

Despite these challenges and concerns in Korean education, it is certain that enthusiasm and diligent efforts attempted by various educational sectors in Korea have led to much progress and development. The passion and support for the development of science and technology has overcome the aftermath of the Korean War and built a modern Korea today. This fact should be respected and celebrated. And, we also need to continuously question the visions of future science education to overcome the current concerns and challenges that the society faces in the presence and future.

References

Driver, R., Leach, J., Millar, R., & Scott, P. (1996). *Young people's images of science. Buckingham*. London: Open University Press.

Hall, G. E., & Hord, S. M. (2011). *Implementing change: Patterns, principles, and potholes* (3rd ed.). Boston: Pearson.

Hodson, D. (2003). Time for action: Science education for an alternative future. *International Journal of Science Education, 25*, 645–670.

Im, S., Yoon, H. G., & Cha, J. (2016). Pre-service science teacher education system in South Korea: Prospects and challenges. *Eurasia Journal of Mathematics, Science & Technology Education, 12*(7), 1863–1880.

Jin, S.-H., & Jang, S. (2007). Elementary school teachers' teaching experience of scientific inquiry. *Journal of Korean Elementary Science Education, 26*(2), 181–191.

Jin, M., & Yoon, H. (2002). *The avoidance of science and engineering fields of high school students in attending colleges and universities in Korea*. Seoul: Korea Research Institute for Vocational Education & Training.

Kang, K. H., & Ahn, K. J. (2014). An analysis of the evaluation and content pedagogical biology questions on the secondary school biology teacher certification examination. *Teacher Education Research, 53*(3), 416–429.

Kim, M., & Tan, A.-L. (2011). Rethinking difficulties of teaching inquiry-based practical work: Stories from elementary pre-service teachers. *International Journal of Science Education, 33*(4), 465–486.

Kim, S., Kim, H.-K., Park, J. H., Jin, E.-N., Lee, M.-J., Kim, S. H., Ahn, Y. K., & Seo, J. H. (2012a). *Trends and international comparative analysis of educational environment in TIMSS*. Seoul: Korea Institute for Curriculum & Evaluation (KICE).

Kim, S., Park, J. H., Kim, H.-K., Jin, E.-N., Lee, M.-J., Kim, J. Y., Ahn, Y. K., & Seo, J. H. (2012b). *Findings from TIMSS for Korea: TIMSS 2011 international results*. Seoul: Korea Institute for Curriculum & Evaluation (KICE).

Kim, M., Tan, A.-L., & Talaue, F. (2013). New vision and challenges in inquiry-based curriculum change in Singapore. *International Journal of Science Education., 35*(2), 289–311.

Kjærnsli, M., & Lie, S. (2011). Students' preference for science careers: International comparisons based on PISA 2006. *International Journal of Science Education, 33*(1), 121–144. https://doi.org/10.1080/09500693.2010.518642.

Kolstø, S. (2001). Scientific literacy for citizenship: Tools for dealing with the science dimension of controversial socioscientific issues. *Science Education, 85*(3), 291–310.

Korea Institute of Curriculum & Evaluation. (2012). *Education in Korea*. Seoul: KICE.

Lee, S. A., Juhn, Y., Hong, J.-E., Shin, Y.-J., Choi, J., & Lee, I. (2007). Difficulties experienced by elementary school teachers in science classes. *Journal of Korean Elementary Science Education, 26*(1), 97–107.

Lee, M.-K., Sohn, W. S., & No, U. K. (2008). *The results from PISA 2006*. Seoul: Korea Institute for Curriculum & Evaluation (KICE).

Lee, B., Shim, K.-C., Shin, M.-K., Kim, J., Choi, J., Park, E., Yoon, J., Kwon, Y., & Kim, Y.-J. (2013). Relationship between science education researchers' views on science educational theories for pre-service science teachers and the examination for appointing secondary school science teachers. *Journal of the Korea Association for Science Education, 33*(4), 826–839.

Liu, E., Liu, C., & Wang, J. (2015). Pre-service science teacher preparation in China: Challenges and promises. *Journal of Science Teacher Education, 26*(1), 29–44. https://doi.org/10.1007/s10972-014-9404-1.

Martin, M. O., Mullis, I. V. S., Foy, P., & Stanco, G. M. (2012). *TIMSS 2011 international results in science*. Chestnut Hill: Boston College.

Ministry of Education, Korea (MOE). (2015). *Science curriculum*. Seoul: Author.

Ministry of Education Science and Technology. (2011). *The second basic plan to foster and support the human resources in science and technology (2011–2015)*. Seoul: MEST.

Na, J., Yoon, H.-G., & Kim, M. (2015). Analysis of the alignment between elementary science curriculum and teacher guidebook: Examining learning objectives in 2009 Grade 3–4 science curriculum. *Journal of Korean Elementary Science Education, 34*(2), 183–193.

OECD. (2014). *OECD science, technology and industry outlook 2014*. Paris: OECD Publishing. https://doi.org/10.1787/sti_outlook-2014-en

OECD (Organisation for Economic Co-operation and Development). (2007). *PISA 2006: Science competencies for tomorrow's world*. Paris: Author.

Osborne, J. (2010). Arguing to learn in science: The role of collaborative, critical discourse. *Science, 328*, 463–466.

Park, S.-J. (2004). A study of science and engineering avoidance situation. *The Journal of Labor and Economy, 27*(1), 56–76.

Park, J. H., Kim, J. Y., & Park, Y. R. (2004). Secondary school science teachers' perceptions of inquiry learning. *Journal of Korean Earth Science Society, 25*(8), 731–738.

Seth, M. (2012). Education zeal, state control and citizenship in South Korea. *Citizenship Studies, 16*(1), 13–28. https://doi.org/10.1080/13621025.2012.651400.

Sim, J.-H., Shin, M.-K., & Lee, S.-K. (2010). Science teachers' perception on major features of the 2007 revised science curriculum for class implementation. *Journal of the Korean Association for Science Education, 30*(1), 140–156.

Yoon, H.-G. (2004). Pre-service elementary teachers' difficulties in science lessons. *Journal of Korean Elementary Science Education, 23*(1), 74–84.

Yoon, H.-G., Kang, N.-H., & Kim, M. (2011). Elementary teachers' conceptions of science inquiry teaching: Cases of South Korea, Singapore and the United States. *Journal of Korean Elementary Science Education, 30*(4), 574–588.

Yoon, H.-G., Kim, M., Kim, B. S., Joung, Y. J., & Park, Y. S. (2013). Pre-service teachers' views of inquiry teaching and their responses to teacher educators' feedback on teaching practice. *Eurasia Journal of Mathematics, Science & Technology Education, 9*(4), 347–359.

Chapter 7
Primary Science Education in Singapore

Yew-Jin Lee

7.1 Overview of Education in Singapore

For nearly two decades, education in Singapore has gained many international accolades that praised its student achievement on international tests such as the Programme for International Student Assessment (PISA) and Trends in International Mathematics and Science Study (TIMSS) (https://www.moe.gov.sg/news/press-releases/equipped%2D%2Dprimed-and-future-ready%2D%2Dsingapore-students-have-what-it-takes-to-thrive-in-the-21st-century-workplace) as well as the quality of its school curriculum (Harlen 1994, Law 2002, Morris and Marsh 1992) and the excellence of its educators (OECD 2010). Teachers here are among the youngest in age in the world and are reported to work just as hard as their counterparts elsewhere in the global North (Lee & Poon, 2014, OECD n.d.), while their students are generally regarded as highly motivated with good reasoning skills (Davie 2012). All of these developments have taken place amidst two ideological pillars enshrined within state education policies since independence in 1965—meritocracy where one can advance in society as far as one is inherently capable of doing (i.e., everyone enjoys an equality of opportunities) and bilingualism, which mandates that every child receives instruction in English and one's mother tongue language to enhance communication and build national identity. As well, there is a high degree of centralization in the administration of schools with nearly every child enrolled in state-run primary schools. Although most jurisdictions might be more than satisfied by these aforementioned outcomes, schooling here has been the subject of constant tinkering in order to improve an already robust education system.

Broad waves of (overlapping) government initiatives over the years have variously emphasized the following: thinking critically, risk-taking, and entrepreneurship

Y.-J. Lee (✉)
National Institute of Education, Nanyang Technological University, Singapore, Singapore
e-mail: yewjin.lee@nie.edu.sg

Y.-J. Lee, J. Tan (eds.), *Primary Science Education in East Asia*, Contemporary Trends and Issues in Science Education 47,
https://doi.org/10.1007/978-3-319-97167-4_7

(e.g., "Thinking Schools, Learning Nation "); mass customization and use of powerful pedagogies that respect different learning styles/needs (e.g., "Ability-driven Education"); holistic, collaborative, and engaged learning coupled with alternative forms of assessment such as project work (e.g., "Teach Less, Learn More"); preparation for life and work in a knowledge-based economy (e.g., "21st Century Competencies"); and cultivating right personal values and love for the nation (e.g., "Character & Citizenship Education"). There has also been a slew of other programs of shorter reach (e.g., the Applied Learning Program to help students make sense of abstract school learning, the life sciences push to enthuse pupils about biotechnology) as well as those that have dealt with structural or technical changes (e.g., the decentralization of school authority, providing more computers in classrooms, widening of teacher career paths, planned road maps for teacher development, etc.). Of no lesser significance have been school-based interventions of various ilks that are almost de rigueur as school leaders try to carve out their distinctive niches in academic, cultural, or sporting fields to attract students. Whether all these innovations and policies have resulted in concrete improvements remains an empirical matter, but it is undeniable that education in Singapore has been in constant evolution at macro-, meso-, and micro-levels (Chan 2011, Lee 2014a).

In so far as education has tried to help local youths fulfill their aspirations, we would be remiss if we left readers believing that there are no challenges or contradictions in teaching and learning in Singapore (e.g., Lee 2008, 2014b). For example, stresses associated with societal expectations about the (extrinsic) value of education as well as exhortations for academic excellence have been widely acknowledged (Chang 2013, Teng 2013). Private tutoring in important school subjects is thus common for those families who can afford it. To a large extent, such pressures to do well in school are direct consequences of the backwash arising from high-stake terminal examinations at Grade 6 (the Primary School Leaving Examination [PSLE]), which determine placement into secondary schools. In recent years, more differentiated or alternative educational pathways have been implemented at the secondary and post-secondary levels of schooling that recognize how talent can indeed be defined and rewarded in multiple ways. By 2021, the four subjects tested in the PSLE will be graded by achievement levels or bands to emphasize learning and individual performance rather than being preoccupied with examination scores. However, entrenched views by parents (and teachers) of what ultimately counts—academic qualifications—are pervasive and are not easily swayed, especially at the primary level of schooling.

According to Hogan et al. (2013), there are two coextensive logics of teaching in Singapore: one that desires more experimental forms of powerful teaching, knowledge building, and student agency, which squarely align with official encouragement for pedagogic creativity, and another that falls back on older, normative elements of teaching and learning within an East Asian moral-psychological paradigm. Unless assessment practices change, no amount of talk about subject matter updating of teachers or the introduction of new teaching methods would gain any purchase if this fundamental assertion is true. Neither would the massive adoption of cutting-edge educational technologies be able to achieve its full potential for learning as was highlighted during an impact evaluation of the Third Masterplan for Information and Communication Technologies in Education (MP3) (Tan et al. n.d.). Whatever is stated in the syllabus will be taught and assessed, thereby maintaining a very tight, mutually reinforcing relationship between national curricular intentions, classroom pedagogy, and student learning outcomes.

7.2 Overview of Primary Science Education and the Curriculum

A comprehensive history of primary science education in Singapore has yet to be written; Tay (1993) provides a rare snapshot of primary science in the early 1990s when it was in a rather tenuous state—fact accumulation by learners was the predominant yardstick of successful science learning contrary to official intent. Poon (2014) and Chin and Poon (2014) have recently sketched how government policies and wider societal events have influenced its expansion (both quantitatively and qualitatively) since 1959. It is therefore hoped that research such as found in this edited book will help to fill the large gaps in our knowledge lest we repeat the mistakes of the past.

Primary education is presently compulsory from Grades 1 to 6 in Singapore and aims to provide a broad range of learning and play activities to develop the full potential in every child (see http://www.moe.gov.sg/education/primary/files/primary-school-education-booklet.pdf). Children with special needs should be similarly covered under this policy from 2018. (There are a very small minority of students that are enrolled in special needs schools or privately run Muslim religious schools or are home-schooled.) Weekly instruction in science ranges from 90 min (Grades 3–4 [lower block]) to over 150 min (Grades 5–6 [upper block]) with the use of ministry-prescribed, commercially produced textbooks and workbooks. Japan and Korea similarly begin the teaching of science at Grade 3, but for different reasons; in Singapore science is delayed to allow learners time to consolidate the learning of two languages and mathematics in the lower grades. Learning English and numeracy from Grade 1 can arguably lay a foundation for the future learning of science. The situation in Mainland China was identical with Singapore, but science now begins in Grade 1 with the onset of recent reforms.

Being one of the four compulsory subjects to be tested in the PSLE (apart from English, mother tongue languages, and mathematics), science (http://www.seab.gov.sg/pages/nationalExaminations/PSLE/2015_PSLE_Subject_Info/0009_2015.pdf) is a valued subject in an economy that has traditionally favored as well as found great wealth from the science, technology, engineering, and mathematics (STEM) industries. (Science, like all other subjects in school other than mother tongue languages and civics, is taught in English.) Provision is also made in years 5 and 6 for students who struggle with any subject to take what are known as foundation courses, which are pegged at more manageable levels of difficulty. Subject-based banding (a form of tracking) in the final 2 years of primary schooling was thus felt to minimize attrition in the subject as well as recognize the wide range of talents that each child possessed. Certainly, these sensibilities are in tandem with a major review of primary school education (the Primary Education Review and Implementation [PERI]) in 2009 (http://www.moe.gov.sg/media/press/files/2009/04/peri-report.pdf) that proposed more engaging instructional methods together with holistic/profile assessments that evaluated the learner as a whole person (e.g., commending instances of effort and progress not just grades) in all the 190 primary schools in the country.

The primary science curriculum is revised every 6 years; the latest iteration came into effect in 2014. Revisions from the 2008 version (CPDD 2007) have not been drastic in nature and involved some removal or addition of learning objectives rather than a recalibration or overhaul of the basic philosophies behind science teaching by the planning committee (Chin and Poon 2014). In line with worldwide trends that have increasingly emphasized science as a highly interesting/fun subject of conducting investigations well within the capabilities of young people (McEneaney 2003), primary science in Singapore has adopted inquiry as *the* defining feature of the curriculum (CPDD 2013). While the teacher is the leader of inquiry, the student is regarded as the key inquirer who raises questions about science in daily life, in his/her environment, and in society. Apart from the learning of scientific content in terms of knowledge and its application, scientific skills and processes are emphasized in the curriculum as well as the cultivation of ethics and attitudes in the discipline.

This spirit of doing science by inquiry was certainly present in earlier versions of the curriculum, but it was only made a defining feature in the 2008 revision. We believe that 2008 was a landmark because inquiry in the curriculum was expressed for the first time in terms of evidence-based reasoning as opposed to mere hands-on or "doing science" with/for kids:

> Teaching science as inquiry must therefore go beyond merely presenting the facts and the outcomes of scientific investigations. Students need to be shown how the products of scientific investigations were derived by scientists and be provided opportunities to: ask questions about knowledge and issues that relate to their daily lives, society and the environment; be actively engaged in the collection and use of evidence; formulate and communicate explanations based on scientific knowledge. (CPDD 2007, p. 11)

Prior to this, for example, in 2001, learning science through process skills seemed to be paramount over the 1980s–1990s in Singapore (Poon 2014), which was very representative of science curricula from the 1960s to 1990s around the world.

Other important learning goals are associated with the learning of the twenty-first-century competencies and the achievement of scientific literacy. Compared to previous eras that made the memorization of scientific facts and definitions paramount, being literate is now considered in terms of being able to reason from evidence and apply knowledge in real-world contexts. This represents a more contemporary version of what it means to be educated in science although this idea was present in some attenuated form in the 2001 version of the curriculum (CPDD 2004).

What was interesting is that in the 2001 version, it was mentioned that one of the aims of an education in science was to "produce competent professionals in the various scientific disciplines who can carry out research and development at the highest level" (CPDD 2004, p. 3). This statement had been removed for unknown reasons by 2008 although such a move recognizes that preparing a minority of learners for future science-related careers is not helpful or socially just for the vast majority of students. The overall aims of primary science in the current 2014 curriculum are thus:

- Provide students with experiences which build on their interest in and stimulate their curiosity about their environment.
- Provide students with basic scientific terms and concepts to help them understand themselves and the world around them.
- Provide students with opportunities to develop skills, habits of mind, and attitudes necessary for scientific inquiry.
- Prepare students toward using scientific knowledge and methods in making personal decisions.
- Help students appreciate how science influences people and the environment (CPDD 2013, p. 5).

In terms of the organization of content, all the intended learning objectives have been grouped (since 2001) into five integrated themes (diversity, interactions, systems, cycles, energy). Although such classifications are somewhat arbitrary and are not to be viewed as watertight compartments, this move was felt to be able to "communicate a more coherent and integrated understanding of science that bridged the life science-physical science divide" (Chin and Poon 2014, p. 33). To a very large extent, this thematic approach and the underlying goal of achieving scientific literacy through student-led inquiry are continued when students study science in Grades 7–8 in Singapore. In common with other centralized school curricula such as Korea, most learning objectives (in the cognitive domain) are located on the lower levels of knowledge and cognitive processes in revised Bloom's taxonomy, which raises issues whether the curriculum could indeed develop deep, critical thinking in science (Lee et al. 2015). Table 7.1 shows the latest version of the coding of the intended primary science learning objectives from Singapore (Lee et al. 2017). (NB. Any profile is necessarily the product of the skill of the coders and does not correspond to the full range of learning experiences offered by any one teacher.)

It was also interesting to note the high degree of alignment for those topics that were in common with the primary science curriculum in Korea, which underscores the extent of similarity in many curricula despite local nuances (Lee et al. 2015). On the other hand, topics in the earth/planetary sciences, human health, rusting, mixtures, dissolving substances, and rainbows among others are absent in the Singapore science curriculum although they are tested in TIMSS, which again provokes speculation about the ideal allocation of time for mastery of facts or the learning of reasoning skills in school (TIMSS 2013). It is therefore an unfounded claim that special provision has been made to prepare local students for these supranational achievement tests given how so many topics are not found in the local curriculum.

The curriculum furthermore encourages teachers to adopt a wide variety of teaching strategies from more teacher-centered ones (e.g., demonstration, direct teaching) to more student-centered types (e.g., games, drama, concept cartoons,

Table 7.1 Table showing learning objectives in the cognitive domain (n = 83) from Singapore classified according to the dimensions of knowledge and cognitive processes in revised Bloom's taxonomy from Lee et al. (2017)

	Remember	Understand	Apply	Analyze	Evaluate	Create	Number of knowledge items
Factual	14 (100) (53.8)	0	0	0	0	0	14 (16.9)
Conceptual	12 (21.4) (46.2)	27 (48.2) (100)	13 (23.2) (50.0)	4 (7.2) (100)	0	0	56 (67.5)
Procedural	0	0	13 (100.0) (50.0)	0	0	0	13 (15.6)
Metacognitive	0	0	0	0	0	0	0
Number of cognitive items	26 (31.4)	27 (32.5)	26 (31.4)	4 (4.7)	0	0	83

Percentages shown in brackets (%)

project work). Instruction also follows a spiral approach where certain concepts (e.g., how plants make food) are revisited in increasing complexity in later years. Above all, inquiry-based learning is supposed to be at the heart of instruction in primary science. To assist teachers conduct and plan for inquiry, the five essential features from the National Research Council (USA) have been described in the official curriculum document as helpful to follow (Loucks-Horsley and Olson 2000). Many teachers here also use the BSCS 5E teaching model of inquiry given its stepwise structure, abundant instructional support, and opportunity for detailed lesson planning.

Everyday assessment in primary science can assume many forms including model-making, posters, debates, portfolios, and practical work apart from the traditional pen-and-paper modes of the PSLE. Again, most teachers in Singapore are to be applauded because they cherish teaching through the hands-on aspects of science and will make special provision to test practical work in school examinations, especially in the upper block.

With regard to science resources, it is fair to conclude that every primary school in Singapore is well equipped with laboratory equipment and curriculum materials/teacher guides supplied or funded by the Ministry of Education (the Curriculum Planning and Development Division). The teaching guide is comprehensive with suggested lesson plans, activities, assessment worksheets, and online resources and is periodically updated. Teachers and other curriculum leaders meet regularly with specialists from the Ministry to share and keep updated on issues pertaining to primary science education. Some schools have provided more resources (e.g., science garden, herb garden, aquaria) and training because they have sought for extra funding from other private/public sources or have strong parent-teacher support groups. Local teachers are also resourceful and share materials with colleagues in the department, while access to other school-wide resources (e.g., computer software, teaching aids) is commonplace. What perhaps remains a major challenge is the scarcity of support technicians to assist teachers in preparing science materials; not every school has these support staff (because it is not mandated) and thus a number of teachers experience logistical difficulties in adequately mounting hands-on activities.

The Singapore Science Center has also been a major institution devoted to the promotion of interest and knowledge of science for the past 40 years. Nearly all school-going children would have had visited it once while in school, and it conducts a wide variety of interesting science activities for both children and adults. Subscription to age-appropriate science magazines and participation in science competitions/fairs of various sorts (national, zonal, school) are other aspects of informal science learning that are available in primary schools.

7.3 Opportunities to Learn for Science Teachers

7.3.1 Initial Teacher Training at the NIE

Nearly every teacher in Singapore has passed through the National Institute of Education (NIE), which is the sole teacher training college in the state. Students enrolled in the undergraduate program (Bachelor of Science [Education] or Bachelor of Arts [Education]) are required to spend 4 years with the NIE with opportunities for field experiences in school and community service projects. Prospective teachers with a university degree require a year of study that incorporates a 10-week teaching practicum before they graduate. In the case of primary school teachers, few are science majors as those with a science background would be typically channeled to teach the subject at higher levels. This postgraduate program has just been extended by 4 months to accommodate an extra teaching practicum given the prevailing emphasis on strengthening theory-practice links and applying knowledge in authentic contexts.

Some of the subject matter method courses that all pre-service primary science teachers at the NIE will study are listed below:

Curriculum and Practices for Primary Science. This course introduces beginning teachers to the ideas behind the nature of science and scientific inquiry, which are the key principles of science teaching and learning. Beginning teachers will also understand the aims, goals, and objectives of the primary science curriculum based on in-depth analyses of the syllabuses, textbooks, and other curricular materials. Relevant general and science-specific educational policies in Singapore will also be discussed. The course also brings in contemporary learning theories and research findings relevant to the teaching practices of primary science.

Assessment in Primary Science. Assessment literacy in primary science will be the central focus in this course. Methods of formal/nonformal assessment as well as the construction and analyses of tests and test results will be examined through hands-on activities. Consideration is also made for assessing practical work in primary science. Beginning teachers will also practice how to provide holistic and helpful feedback for pupils through verbal and written means.

Pedagogies for Primary Science. This course introduces beginning teachers to innovative ways of teaching and learning science both within and outside the classroom, especially through the use of collaborative inquiry activities. They will also have opportunities to critically practice and reflect on their teaching skills in the areas of planning, instruction, and assessing for understanding during microteaching lessons.

Meeting Learners' Needs in Primary Science. Engaging teaching methods that align with the needs of 21CC learners will be the learning goal of this course. Beginning teachers will also examine their personal philosophies of science teaching so that they can learn how to balance their teaching of knowledge with skills and values. Catering to the needs of diverse learners through differentiating instruction and the giving of holistic feedback assessment will also be practiced (see http://www.nie.edu.sg/study-nie/admissions/teacher-education-undergraduate-studies/postgraduate-diploma-education-programmes).

It has been suggested as long ago as 1998 that teacher training and local classroom processes were important contributors toward high national achievement in TIMSS. Reasons include an "emphasis…on pragmatic and instructional issues rather than social and equity issues….less on the analysis of theoretical constructs and more on the application of concepts and practices" (Wong et al. 1998, p. 37). Indeed, critical theories of learning and cultural/social justice aspects of science instruction are intentionally kept to a minimum at this stage although these topics are taught at the master's or doctoral level at the NIE. Wong et al. (1998, p. 37) elaborate how the "closeness of fit between teacher preparation and school practice ensures that beginning teachers are adequately prepared for their teaching," which we believe will experience even greater correspondence with the impending reforms to lengthen teacher training in NIE.

7.3.2 Professional Development (PD) and Higher Degrees at the NIE

Being a young workforce with an average age of 36 years, teacher training exerts a strong positive influence on teachers' professional behaviors and knowledge (OECD, n.d.). On average, 98% of teachers have participated in professional development (PD) activities, 61% in education conferences, and 53% in teacher network activities, which all exceeded OECD norms in 2013. Even though Lee and Poon (2014) have cautioned about the large variability in the time spent on various work activities among teachers, it has been reported that local teachers devote more time to lesson preparation and marking compared to others in OECD countries (OECD n.d.). It is fair to assume that such generic indicators are likewise applicable to local primary science teachers although specific data are absent.

In common with other nations, many primary science teachers are not science majors, and hence they often struggle in the teaching of content. To address this problem, the Ministry of Education is ramping up and deploying teachers who are specialists in the field such as in science (http://www.moe.gov.sg/media/press/2014/09/growing-our-teachers-building-our-nation.php). These experts will therefore complement the Subject Heads and Heads of Department in science (http://www.moe.gov.sg/media/press/1996/st00196.htm). A number of content upgrading, formal certification, and PD courses (see below) are provided at the NIE, but the bulk of their learning is normally derived from learning on the job as well as from career-long PD courses from the Academy of Singapore Teachers (AST), Ministry, peer-sharing, or communities of practice (department, school, cluster, zonal), among others. Moreover, the degree and spontaneity of support provided by the Ministry curriculum specialists when any teacher encounters difficulties or questions regarding the implementation of the national curriculum is truly exceptional.

Teachers who enroll in the NIE Advanced Diploma in Primary Science Education program (about 2-year duration on a part-time basis) may then join the bachelors' program at the NIE upon successful completion. Courses in the Advanced Diploma include content learning in the biological and physical sciences, introduction to the thematic structure of the curriculum, assessment and testing issues, and modules devoted to enhancing the pedagogical content knowledge and inquiry skills to effectively teach the five themes in the primary science curriculum (http://www.nie.edu.sg/studynie/professional-development-courses/programmes-courses/advanced-diploma-programmes/advanced-d-5). The Certificate in Primary Science Education is a less intensive course (about 1.5-year duration on a part-time basis) with two modules revisiting subject matter knowledge for primary levels and another module devoted to assessment theory and practices (http://www.nie.edu.sg/study-nie/admissions/graduate-studies-professional-learning/professional-development-programmes-and-courses/certificate-2).

A large variety of ad hoc PD courses at the NIE are available and encompass STEM education, scientific argumentation, discourse studies in science, case-based science investigations, problem-based learning, modelling instruction, as well as topics in the natural sciences (e.g., plant physiology, animal behavior, green chemistry, ecology) (http://www.nie.edu.sg/study-nie/admissions/graduate-studies-professional-learning/professional-development-programmes-and-courses/courses/ipd0121). These short courses (NB. topics can vary over time) are open to all teachers who then incorporate them into their personal PD plans for the school year given that each teacher in Singapore is entitled up to 100 h of paid PD leave.

Three categories of higher degree programs at the NIE are available for teachers who wish to upgrade themselves: higher degrees by coursework (MSc, MEd), master's and PhD by research, and the Doctorate in Education (EdD). Mention must also be made of the popular Enhanced Professional Development Continuum Model since 2012 whereby graduate teachers are sponsored by the Ministry for a 2- to 4-year master's degree in multiple disciplines, including science education. In the latter, instruction would be provided in curriculum studies, critical/cultural studies, research methods and methodology, scientific argumentation, inquiry studies, and representations, among others (http://www.nie.edu.sg/study-nie/admissions/graduate-studies-professional-learning/higher-degree-programmes/masters-coursework/master-education/science). This program has graduated many teachers of science, and it is hoped that they have been empowered in various ways through this course especially through the capstone research project performed by the more able students. Past research topics in primary science here have included inquiry science, cogenerative dialogues, assessment for learning, Toulmin's argument patterns, problem-based learning, children's misconceptions, questioning techniques, and the application of refutation texts, among many others.

7.3.3 Professional Development at the AST

The Academy of Singapore Teachers (AST), which is run by the Ministry of Education, further provides a plethora of PD for all teachers across subjects, which their mission statement captures well—"building a teacher-led culture of professional excellence centred on the holistic development of the child" (AST 2012). As well, there is a subject chapter in primary science led by master teachers who are experts in classroom pedagogy. Based on the theory of networked learning, the purposes of the subject chapter are to "equip Primary Science teachers with impactful and innovative pedagogies that are underpinned by sound learning theories, experimentation and research. It is a professional network to raise professional standard in science disciplines in general and primary in particular" (http://www.academyofsingaporeteachers.moe.gov.sg/networked-learning-communities/subject-chapters/primary-science-chapter).

Past courses organized by the subject chapter have included learning journeys to science-rich outdoor sites (e.g., Fort Canning Park, the HortPark garden), how to conduct classroom research, and ways to enhance the pedagogical content knowledge for teaching primary science. One example of the frequently deep and rich conversations among these practicing teachers who attended these courses resulted in an action research project on comment-only marking strategies as part of assessment for learning (AfL). As how this teacher articulated her feelings, it was literally a life-changing professional learning moment for her as she realized the power of teacher feedback (https://tinyurl.com/mbfw6kl). And because scientific argumentation has caught the imagination of many local primary science teachers in Singapore, another group of teachers decided to embark on an action research project to see how learning the components of argumentation could help their students answer examination questions with greater accuracy and thoroughness. Pleased with the results of their intervention, these teachers then shared their findings with the rest of the school and persuaded curriculum leaders to implement it as a school-wide learning strategy (https://tinyurl.com/lo8s8t9).

7.3.4 Science Education Conferences

All science teachers in Singapore have the opportunity to attend the International Science Education Conference (ISEC) (http://www.isec2014.org/) held at the NIE once every 3 years. In addition, we have the Singapore International Science

Teachers' Conference organized biannually by the AST and other organizations such as the Science Teachers' Association of Singapore, Curriculum Planning and Development Division of MOE, Singapore Science Center, and Academy of Principals (Singapore) (https://www.sistc.sg/). On occasion, teachers can avail themselves to workshops and talks or guest lectures by visiting prominent researchers. The latter has increased over the years since the education system in Singapore has been in the world's limelight.

It is clear by now that capacity building is both abundant and frequent for primary science teacher in Singapore. For example, the Educational Technology Division of the Ministry has organized a number of activities anchored in the area of primary science. In such intentionally collaborative projects, education technology specialists and professional researchers work closely with school teachers in small teams on specific research topics such as Afl or argumentation and the use of ICT during science (see https://sites.google.com/a/moe.gov.sg/igps-afl-2015/). Some experienced teachers have also turned the tables around and conducted short workshops for parents on how they could support their children in the learning of science by inquiry. These well-attended sessions give adults a chance to practice mini-investigations and thereby better understand the process of thinking scientifically.

7.4 Primary Science Education Research in Singapore

It is most gratifying that primary school teachers, not just in science, are the most receptive toward the introduction and field-testing of educational innovations in Singapore. Research access often becomes a matter of formality as classroom practitioners in this period of schooling are open to trying new activities to make the learning of science more appealing. At the same time, we will show later that there is still room for improvement in how actual lessons are conducted and what types of knowledge our students routinely handle as the system aspires toward learning in/for the twenty-first century.

7.4.1 Informal Science Learning Environments

Informal science learning environments represent an exciting yet somewhat baffling terrain for local primary science teachers even though the PERI review has recommended more outdoor learning science facilities, learning journeys (i.e., field trips), and laboratories. Caleon and Subramaniam (2008) report that upper primary students attending a cryogenics enrichment course at the Singapore Science Center had increased enjoyment and interest in pursuing science-related careers though there was no change in their views toward the social implications of science. Anchoring their research in an outdoor nature learning camp, Yeo and Lee (2012) show how personal interest in nature was the vital trigger that allowed students from Grades 3 to 5 to pursue improvable questions/ideas following a scaffolded knowledge building pedagogy. For some pupils, their motivation in wanting to discover answers to their investigations about rotting meat had led them to learn about fair testing and abstract ideas about abiogenesis. Nonetheless, such successful cases of learning were few and mitigated by the immediate contexts of the school. That is, how their teachers facilitated these open-ended forms of inquiry largely depended on what the teachers believed was actually inquiry in the first instance—knowledge building was in fact a boundary object accommodated conflicting interpretations of the same phenomenon/event (Yeo 2014).

7.4.2 Teachers of Primary Science

Another teacher described by Tan and Wong (2012) exemplifies our previous contention about how teachers powerfully mediate learning: Lynn lost opportunities to engage meaningfully with a group of students who had obtained unexpected results during an experiment. She lost a valuable moment to boost her students' epistemic agency because it was not in her lesson agenda. In a post-lesson conference, Lynn admitted that getting her students to know the "correct answers" and coverage/completion of content was a priority, not the encouragement of student inquiry learning per se. Lest these examples give a wrong impression of local science teachers, Lee (2008) explained how a teacher practiced guerilla science teaching to camouflage her early advocacy of inquiry science. Helen was exceptional in freeing up hybrid spaces for students to take charge of their own learning as much as she was faithful to her own philosophies of science learning. One could, of course, legitimately criticize Helen's colleagues for being overly dismissive of inquiry science as a viable teaching method at that point in time. Other elementary science teachers have been bold in attempting to import powerful but difficult teaching methods into their classrooms such as knowledge building (e.g. Lee 2010b), advanced questioning techniques and concept cartoons (Han 2011), use of puppets and argumentation (Sim 2011), and the use of cases to stimulate thinking (Chong 2012). From related work by Chin T. Y. (2013), it was found that the productive questioning strategies routinely employed by primary science teachers mimicked those by secondary science teachers (see Chin 2007) although we are also told that the narrow/closed initiation-reply-evaluate (IRE) format was dominant in classroom discourse in Singapore (Lossman and So 2010).

7.4.3 Students of Primary Science

Researchers have tended to neglect hearing the voices of students; hence, studies that report their perspectives of teaching and learning reforms are significant. Based on ethnographic observations, Alice was a Grade 4 student who strategically calculated and weighed her chances of getting ahead in her science class (Lee, Y.-J. 2010). Compared to her earlier self in the school year, she slowly deidentified herself as a keen learner over the months due to a string of negative episodes that turned her away from science. As was described earlier, Alice, like most other students and parents, knew that academic prowess was vitally important.

Other Grade 4 students described by Zhai et al. (2014) drew images of science and how they saw themselves as learners through a Draw-A-Scientist task. While 60% drew themselves doing hands-on activities, 20% depicted students passively listening to or watching demonstrations from a distance by the teacher. The most common descriptors of everyday science learning used by students were "'doing science as conducting hands-on investigations," "doing science as learning from the teacher," "doing science as completing the workbook," and "doing science as a social process." Students additionally said that they needed to be well behaved; their job was mainly to listen and follow instructions that severely minimized the minds-on components of inquiry, an issue that persists even at secondary levels in Singapore (Chue and Lee 2013). Whether in Singapore or elsewhere, we know far too little about educational change from the perspective of the child because it is almost a standard practice to treat them merely as the objects of the research to be "worked on" by the experts.

7.4.4 Inquiry-Based Learning

From a 14-month observational study of four primary school teachers doing inquiry science, it was found that they were very conscious of classroom contexts, specifically the learning needs of every pupil (Poon et al. 2012). While they were cognizant of the theoretical aspects of learning science by inquiry, they managed to blend in or hybridize what their many years of experience had taught them about what makes for an effective science lesson. In particular, they ensured that (1) students knew the cognitive and procedural tasks before performing an investigation, (2) they repeated certain components of inquiry when students failed to understand through their continual monitoring of student learning, and (3) there was synthesis of concepts at the consolidation phase of the lesson. As the authors have summarized, "Apart from supporting the prevalence of the core components of inquiry found in theoretically-driven pedagogical frameworks, it revealed nuances of practice that were regarded as integral to the success of day-to-day enactments of inquiry in Singapore" (Poon et al. 2012, p. 324).

Another research study into the practice of inquiry teaching in Singapore was conducted by Kim et al. (2013). They found that many local teachers confounded the meaning of inquiry science when they privileged the learning of content over students' attempts at explanations, discussions, presentation, and communication of their knowledge. Although both pre-service and in-service teachers in their sample saw themselves as facilitators of inquiry, the former highlighted the learning of process skills, whereas the latter valued knowledge acquisition during inquiry lessons.

The in-service teachers were understandably more pragmatic and were painfully aware of their institutional responsibilities to prepare students for the examinations. This inescapable social fact thus dampened the overall enthusiasm and desire to expand opportunities to learn science by inquiry for both groups of teachers.

7.4.5 Types of Knowledge Enactment

What types of knowledge do primary school students encounter and handle in their classrooms? What knowledge of the discipline of science and of pedagogy in general do teachers possess? How do teachers teach in Singapore? A decade-old study by Hogan and Gopinathan (2008) offered some preliminary clues concerning these vital questions about primary school teachers in general. Table 7.2 shows the types of knowledge structures (i.e., the concept of disciplinarity) found during the implemented/enacted curriculum in the aggregated subject areas of English, mathematics, science, social studies, and mother tongue. These findings reveal that teaching had limited disciplinarity; authoritative knowledge was chiefly transmitted from the teacher or textbooks to be learnt (passively) by the pupil. The most traded forms of knowledge, moreover, were located at the lower ends of the spectrum (i.e., factual level) instead of more advanced/linked concepts or procedural (knowing how) knowledge.

In terms of their choice of teaching methods or how they organized tasks, the authors report that teacher-dominated pedagogy was extremely common in primary (and secondary) classrooms, at least in Grade 5. Using a unit of analysis known as a "phase" (i.e., a sustained activity for more than 5 min), Table 7.3 shows the results of their codings from 591 lessons in the same subject areas of English, mathematics, science, social studies, and mother tongue languages.

Table 7.2 Table showing the types of knowledge or disciplinarity (coded from 0 to 3) observed in Grade 5 classrooms in Singapore adapted from Hogan and Gopinathan (2008)

Knowledge classification	Score in Grade 5 classrooms (based on levels 0–3)
Depth of knowledge	
Basic/fact/rote	2.4
Procedural/how-to	0.7
Advanced concept	0.1
Knowledge manipulation	
Reproduction	2.3
Application	0.4
Generation of knowledge new to students	0.1
Knowledge assessment	
Truth—a function of what teacher/text says	2.6
Comparison	0.2
Critique	0.1

Data was obtained from classroom observations of a random sample of 1128 lessons in 2004/2005 by the authors

Table 7.3 Table showing the overall classroom organization of Grade 5 lessons adapted from Hogan and Gopinathan (2008)

Types of activities in Grade 5 classrooms (rank ordered)	% of phases
Class answer checking/IRE	27.6
Whole class lecture/monologue	20.5
Individual seatwork	18.3
Small group work	12.1
Whole class elicitation and discussion	6.3
Student demonstration	5.2
Whole class demonstration/activity	3.3
Choral repetition	3.0
Laboratory/experiment	2.3
Test-taking	1.6

Finally, the authors state that content knowledge and pedagogical content knowledge among primary school teachers (through self-reports) were the highest for mathematics ($M_{CK} = 4.93$, $M_{PCK} = 4.48$, from levels 1 to 6) followed by science ($M_{CK} = 4.15$, $M_{PCK} = 3.96$) and English ($M_{CK} = 4.01$, $M_{PCK} = 3.89$). Taking these results together, teaching a decade ago definitely assumed a more traditional didactic and controlling stance where student epistemic agency was minimized although we are uncertain if this situation was true for science classrooms. These are truly interesting findings and raise more questions than answers about how pedagogy and knowledge are mediated in Singapore.

7.4.6 Everyday Classroom Practices

A final area that is perhaps severely lacking here with respect to publications from evidence-based knowledge about local primary science, and science education in general, are large-scale, longitudinal studies into teaching and learning. These panel studies have proven to be valuable in unpacking the complex and contradictory subtleties of "everyday" schooling of the kind described by Hogan and Gopinathan (2008). The small corpus of primary science education research in Singapore thus far has been overwhelming confined to data coming from a few classrooms or limited sample sizes of participants as is typical of qualitative research. A recent baseline panel study involving video data from five Grade 5 teachers who each taught a unit of the science curriculum showed some of the following interesting features:

- Majority of classroom talk was factual talk (0.58 out of a unit of 1.0) followed by explanatory talk (0.42) and procedural talk (0.38). Accordingly the knowledge focus of talk was factual knowledge (0.59), conceptual knowledge (0.49), and procedural knowledge (0.39). Handling of epistemic and metacognitive knowledge was uncommon.
- Teacher questions were overwhelmingly of the closed type (0.88).

- Classrooms were dominated by teacher talk (and sometimes demonstrations) with students listening to the teacher, answering teacher questions, and taking notes (0.69). Checking or reviewing work or revising occupied 0.34 of the time, the next highest category.
- Teacher-directed inquiry was predominant with teachers emphasizing concept development, not just memorizing facts. Teachers felt that promoting or maintaining interest in science was very important as was making science relevant to daily life. Teachers felt that examination and answering techniques were critical to help students move to the next level (Kwek et al. 2017).

7.5 Closing Thoughts

We have much to be appreciative about the state of primary science education in Singapore: The curriculum is believed to be coherent and aligned with assessment practices; teaching is typically thorough in nature and has pockets of experimentation and creativity spearheaded by a highly trained cadre of committed teachers; schools here are well-stocked with laboratory equipment and other material benefits, and students enjoy the by-products of studying in one of the most technologically connected and prosperous societies on earth. What should not be forgotten is that many of these developments did not occur overnight but instead were in all likelihood a series of (beneficial, or at least neutral) refinements and tweaks interacting with an economy that was on an upward trend (Gopinathan 1997, Hanushek and Woessmann 2015). Institutional cooperation and close personal ties between staff from the NIE, AST, and Ministry of Education have further ensured that teacher training and PD are kept in tandem with the latest policy mandates and research findings from abroad (c.f. Chin and Poon 2014). These are again some of the tangible outcomes of working in a small island state that has enshrined human capital development as its central dogma. It does not take long for any visitor to realize that taking a stance of "business as usual" in Singapore is highly unusual, especially when it comes to the task of educating the young.

Recall that we made the claim that assessment or examination practices appeared to be a bottleneck that is preventing us from fully realizing much of the intended effects of school reforms in Singapore. What we are now witnessing is a gradual shift toward more creative or novel ways of asking questions in the PSLE whereby students have to apply their knowledge in unfamiliar contexts (see http://www.straitstimes.com/singapore/interesting-tricky-psle-science-paper-scores-well-with-parents). For too long, teachers have wanted policymakers to expand the range of assessment modes accepted in the PSLE (Lee 2010a). If these radicalizing trends continue, they will filter down to the classroom, which surely augurs well for elevating primary science education to an even higher level through lessening the culture of performativity here (Leong and Tan 2014).

Our guarded optimism, however, does not ignore the bigger if not more important existential purposes of education and the development of personhood to which an education in science ought to be legitimately subsumed (Tan 2013). These difficult yet immensely consequential inquiries have to be raised and debated often in the public sphere, which has not happened to any substantial degree here. It is indeed likely that too many reforms in Singapore have overemphasized the technical-rational aspects of instruction at the expense of the fundamental "what" and "why" questions (Teo et al. 2013), which is to say that the means and not the ends of education have been privileged. Nor has there been enough consultations with arguably the most critical group of stakeholders in primary education—the young people of Singapore. Their voices and interests must be heard more often and taken more intently because learning STEM has both theoretical and instrumental application in adulthood (Rudolph 2005). Speaking from an interdisciplinary curricular perspective, greater attention must also be paid to learning about climate change and encouraging forms of planetary ethics, practical issues of the kind that surely can determine the longevity or extinction of humankind.

Little, if no, research has been done on the impact of socioeconomic background on science achievement in Singapore. In view of the growing official and societal concern over educational inequalities, this topic merits further attention. In addition, the different attainment levels in primary science achievement between foundation and non-foundation bands may have repercussions for students' achievement in secondary school science. For example, students in foundation science forego learning topics such as forces in springs, cells, and energy conversion apart from having about half the number of open-ended and multiple-choice questions (with just three options) with six to seven structured questions (e.g., fill-in-the-blanks, matching) that are absent in the standard track (http://www.seab.gov.sg/content/syllabus/PSLE/2016_PSLE_Subject_info/0039_2016.pdf).

References

AST [Academy of Singapore Teachers]. (2012). *Our mission and vision*. Retrieved from http://www.academyofsingaporeteachers.moe.gov.sg/about-ast/our-mission-and-vision

Caleon, I., & Subramaniam, R. (2008). Attitudes towards science of intellectually gifted and mainstream upper primary students in Singapore. *Journal of Research in Science Teaching, 45*, 940–954.

Chan, C. M. L. (2011). *An analysis of the 2004/05 Teach Less, Learn More policy in Singapore*. Unpublished doctoral dissertation, The University of Western Australia, Perth, Australia.

Chang, R. (2013, September 7). Faster rat race? No stressful schooling? No more foreigners? No. *The Straits Times*, p. D2.

Chin, C. (2007). Teacher questioning in science classrooms: Approaches that stimulate productive thinking. *Journal of Research in Science Teaching, 44*, 815–843.

Chin, T. Y. (2013). *Insights and lessons from teachers' initial professional learning and collaborative practices in questioning for higher-order scientific discourse in primary science classrooms*. Unpublished education doctorate thesis, National Institute of Education, Singapore.

Chin, T.-Y., & Poon, C. L. (2014). Design and implementation of the national primary science curriculum: A partnership approach in Singapore. In A.-L. Tan, C. L. Poon, & S. S. L. Lim (Eds.), *Inquiry into the Singapore science classroom: Research and practices* (pp. 27–46). Dordrecht: Springer.

Chong, C. G. A. L. (2012). *Using cases to engage and stimulate students' thinking in science.* Unpublished masters' thesis, National Institute of Education, Singapore.

Chue, S., & Lee, Y.-J. (2013). The proof of the pudding?: A case study of an "at-risk" design-based inquiry science curriculum. *Research in Science Education, 43*, 2431–2454.

CPDD [Curriculum Planning & Development Division]. (2004). *Science syllabus primary 2001.* CPDD: Ministry of Education, Singapore. Retrieved from http://www.moe.gov.sg/education/syllabuses/sciences/files/science-primary-2001.pdf

CPDD. (2007). *Science syllabus primary 2008*. CPDD: Ministry of Education, Singapore. Retrieved from http://www.moe.gov.sg/education/syllabuses/sciences/files/science-primary-2008.pdf

CPDD. (2013). *Science syllabus primary 2014*. CPDD: Ministry of Education, Singapore. Retrieved from http://www.moe.gov.sg/education/syllabuses/sciences/files/science-primary-2014.pdf

Davie, S. (2012, December 12). S'pore pupils top tests, score higher in reasoning. *The Straits Times*, p. A1.

Gopinathan, S. (1997). Educational development in Singapore connecting the national, regional and the global. *The Australian Educational Researcher, 24*, 1–12.

Han, S. T. K. (2011). *Effecting conceptual change and development through the optimal use of concept cartoons: Student argumentation, questioning and cooperative learning at play.* Unpublished masters' thesis, National Institute of Education, Singapore.

Hanushek, E. A., & Woessmann, L. (2015). *The knowledge capital of nations: Education and the economics of growth.* Cambridge, MA: The MIT Press.

Harlen, W. (1994). Primary science. In T. Husén & T. N. Postlethwaite (Eds.), *The international encyclopedia of education* (pp. 5328–5335). Oxford: Pergamon.

Hogan, D., & Gopinathan, S. (2008). Knowledge management, sustainable innovation, and pre-service teacher education in Singapore. *Teachers and Teaching, 14*, 369–384.

Hogan, D., Chan, M., Rahim, R., Kwek, D., Khin, M. A., Loo, S. C., Sheng, Y. Z., & Luo, W. (2013). Assessment and the logic of instructional practice in secondary 3 English and mathematics classrooms in Singapore. *Review of Education, 1*, 57–106. https://doi.org/10.1002/rev3.3002.

Kim, M., Tan, A.-L., & Talaue, F. (2013). New vision and challenges in inquiry-based curriculum change in Singapore. *International Journal of Science Education, 35*, 289–311.

Kwek, D., Lee, Y.-J., & Wong, H. M. (2017). *Initial findings of core research programme: Baseline investigations of science pedagogy*. Presentation at AST science seminar on 21 March 2017 at the Academy of Singapore teachers, MOE, Singapore.

Law, N. (2002). TIMMS science results for Hong Kong: An ailing dragon with the British legacy. In D. F. Robitaille & A. E. Beaton (Eds.), *Secondary analysis of the TIMMSS data* (pp. 157–175). Dordrecht: Kluwer Academic Publishers.

Lee, Y.-J. (2008). Thriving in-between the cracks: Deleuze and guerrilla science teaching in Singapore. *Cultural Studies of Science Education, 3*, 917–935.

Lee, I. (2010). *Assessment for contemporary needs: The need for realignment*. Saarbrücken: VDM Verlag Dr. Müller.

Lee, Y.-J. (2010). Not if but when pedagogy collides with culture in Singapore. *Pedagogies: An International Journal, 5*, 17–26.

Lee, Y.-J. (2014a). Knowledge creation in the mangle of practice: Implications for educators. In S. C. Tan, H.-J. So, & J. Yeo (Eds.), *Knowledge creation in education* (pp. 167–188). Dordrecht: Springer.

Lee, Y.-J. (2014b). Science education in a straightjacket: The interplay of people, policies, and place in an east-Asian developmental state. In A.-L. Tan, C. L. Poon, & S. S. L. Lim (Eds.), *Inquiry into the Singapore science classroom: Research and practices* (pp. 165–189). Dordrecht: Springer.

Lee, Y.-J., & Poon, C. L. (2014). The professional work of teachers in Singapore: Findings from a work-shadowing study. *Asia Pacific Education Review, 15*, 525–535.

Lee, Y.-J., Kim, M., & Yoon, H.-G. (2015). The intellectual demands of the intended primary science curriculum in Korea and Singapore: An analysis based on revised Bloom's taxonomy. *International Journal of Science Education, 37*, 2193–2213.

Lee, Y.-J., Kim, M., Jin, Q., Yoon, H.-G., & Matsubara, K. (2017). *East-Asian primary science curricula: An overview using revised Bloom's taxonomy*. Dordrecht: Springer.

Leong, W. S., & Tan, K. (2014). What (more) can, and should, assessment do for learning? Observations from 'successful learning context' in Singapore. *The Curriculum Journal, 25*, 593–619.

Lossman, H., & So, H. J. (2010). Toward pervasive knowledge building discourse: Analyzing online and offline discourses of primary science learning in Singapore. *Asia Pacific Education Review, 11*, 121–129.

Loucks-Horsley, S., & Olson, S. (Eds.). (2000). *Inquiry and the national science education standards: A guide for teaching and learning*. Washington, DC: National Academies Press.

McEneaney, E. H. (2003). Elements of a contemporary primary school science. In G. S. Drori, J. W. Meyer, F. O. Ramirez, & E. Schofer (Eds.), *Science in the modern world polity: Institutionalization and globalization* (pp. 136–154). Stanford: Stanford University Press.

Morris, P., & Marsh, C. (1992). Curriculum patterns and issues in East Asia: A comparative survey of seven East Asian societies. *Journal of Education Policy, 7*, 251–266.

OECD. (n.d.). Singapore: Country note: Results from 2013. Retrieved from http://www.oecd.org/edu/school/TALIS-2013-country-note-Singapore.pdf

OECD [Organisation for Economic Co-operation and Development]. (2010). Singapore: Rapid improvement followed by strong performance. Retrieved from http://www.oecd.org/countries/singapore/46581101.pdf

Poon, C. L. (2014). Five decades of science education in Singapore. In A.-L. Tan, C. L. Poon, & S. S. L. Lim (Eds.), *Inquiry into the Singapore science classroom: Research and practices* (pp. 1–26). Dordrecht: Springer.

Poon, C. L., Lee, Y.-J., Tan, A.-L., & Lim, S. S. L. (2012). Knowing inquiry as practice and theory: Developing a pedagogical framework with elementary school teachers. *Research in Science Education, 42*, 303–327.

Rudolph, J. L. (2005). Inquiry, instrumentalism, and the public understanding of science. *Science Education, 89*, 803–821.

Sim, L. Y. C. S. (2011). *Puppets for science talk: Its use and impact on students' talk in primary science in the Singapore context*. Unpublished masters' thesis, National Institute of Education, Singapore.

Tan, J. (2013). Aims of schooling for the twenty-first century: The desired outcomes of education. In Z. Deng, S. Gopinathan, & C. K.-E. Lee (Eds.), *Globalisation and the Singapore curriculum* (pp. 33–47). Dordrecht: Springer.

Tan, A.-L., & Wong, H.-M. (2012). "Didn't get expected answer, rectify it.": Teaching science content in an elementary science classroom using hands-on activities. *International Journal of Science Education, 34*, 197–222,

Tan, S. C., Chai, C. S., Lee, C. B., Teo, K. G. T., Chen, W., Koh, J. H. L. et al. (n.d.). *Evaluation of implementation of the IT Masterplan 3 and its impact on Singapore schools: Instrumentation and baseline study*. Retrieved from https://www.nie.edu.sg/files/NIE_research_brief_11-001.pdf

Tay, E. L. K. (1993). *The teaching and learning of primary science in Singapore*. Unpublished doctoral thesis, Monash University, Melbourne, Australia.

Teng, A. (2013). '90 marks is never enough'. *The Straits Times*, p. B1.

Teo, J. E., Deng, Z., Lee, C. K.-E., & Lim-Ratnam, C. (2013). Teach less, learn more: Lost in translation. In Z. Deng, S. Gopinathan, & C. K.-E. Lee (Eds.), *Globalisation and the Singapore curriculum* (pp. 99–117). Dordrecht: Springer.

TIMSS [Trends in International Mathematics and Science Study]. (2013). *TIMSS2011 Frameworks – Chapter 2*. Retrieved from http://timss.bc.edu/TIMSS2011/downloads/TIMSS2011_Frameworks-Chapter2.pdf

Wong, A. F. L., Goh, S. C., Gopinathan, S., & DeRozario, V. (1998). Singapore's primary schools. *Education 3–13, 26*, 33–41.

Yeo, J. (2014). Knowledge building as a boundary object in formal/informal learning. In A.-L. Tan, C. L. Poon, & S. S. L. Lim (Eds.), *Inquiry into the Singapore science classroom: Research and practices* (pp. 237–249). Dordrecht: Springer.

Yeo, J., & Lee, Y.-J. (2012). Knowledge advancement in environmental science through knowledge building. In D. K. C. Tan & M. Kim (Eds.), *Issues and challenges in science education research: Moving forward* (pp. 317–322). Dordrecht: Springer.

Zhai, J., Jocz, J., & Tan, A.-L. (2014). "Am I like a scientist?": Primary children's images of doing science in school. *International Journal of Science Education, 36*, 553–576.

Part II
The Expert Commentaries

Chapter 8
The End of Science Education in East Asia?

Adam Lefstein

The title of this chapter is intended as homage to Neil Postman's (1995) *The End of Education*, which though written over two decades ago, on the other side of the globe, resonates with the chapters of this book. Postman decries the focus in US educational discourse on *technical* questions – such as how to train teachers, which instructional strategies work best, or how to construct assessments – and the concomitant neglect of *metaphysical* questions about what education is for. "End" is of course polysemous, meaning both purpose (as in, "the ends of education") and finish (as in, "the end of schooling as we know it"). Postman intends both meanings, suggesting that if we cannot provide students, teachers, and the public a convincing answer to the metaphysical question about the ends of education, then public schooling will be finished.

Postman casts this metaphysical question in terms of the stories we tell ourselves about education – "not any kind of story, but one that tells of origins and envisions a future, a story that constructs ideals, prescribes rules of conduct, provides a source of authority, and, above all gives a sense of continuity and purpose." Such stories give meaning to the educational endeavor and inevitably also shape the answers that we give to the technical questions as well. Following Postman's lead, I examine here the metaphysics of science education in the East Asian countries discussed in the book through analysis of the stories offered in the chapters. I'm interested in both the stories the authors narrate – their own stories, and those of the policy-makers and public they describe – and, reading between the lines, the stories that are conspicuously absent.

The dominant story, which most frequently recurs in the chapters, is the *global competitiveness* story. Wu, Wang, Cheng, and Yang open the Taiwan chapter with this story: "National competitiveness and economic growth of Taiwan mainly relies on the development of its science and technology industries." Similar statements

A. Lefstein (✉)
Ben-Gurion University of the Negev, Be'er Sheba, Israel

Y.-J. Lee, J. Tan (eds.), *Primary Science Education in East Asia*, Contemporary Trends and Issues in Science Education 47,
https://doi.org/10.1007/978-3-319-97167-4_8

(though not all so compact) appear in the chapters on Korea, Japan, Hong Kong, and Singapore. The basic plot of this story is that (a) the global knowledge economy is changing rapidly; (b) to compete in this economy requires a scientifically and technologically adept, flexible, and innovative workforce; and (c) therefore, science and technology education are critical to national prosperity. Many of the chapters also emphasize the importance to their economies of human capital in light of their country's lack of natural resources.

The global market rewards competitive advantage, so it is not sufficient to be good; each country needs to be better than all the rest (i.e., it's a zero-sum game). As such, even the high performers appearing in this book display some insecurity about their capacity to continue to maintain their competitive edge. Wu and colleagues end the Taiwan chapter with the concern that, though Taiwan scored among the top ten in TIMSS 2011 achievement tests, their fourth graders scored below the top ten in interest and confidence in learning science. The authors call this outcome "a crisis but also a turning point", i.e., an opportunity to improve the quality of primary science education in Taiwan in order "to be one of the top-rank countries in the world."

Postman warns about the US version of the global competitiveness story, which he labels economic utility, in *The End of Education*. First, he points to the lack of evidence tying national productivity to educational quality – a point which is hotly debated among economists (see, e.g., Hanushek and Woessmann 2016, and, for a critique, Komatsu and Rappleye 2017). Writing from Israel, the so-called start-up nation, Postman's assessment rings true: our current high-tech boom is more likely related to cultural, economic, and geopolitical factors than our mundane school science (we ranked 40th in PISA 2015 Science). Second, more fundamentally, Postman argues that "any education that is mainly about economic utility is far too limited to be useful, and, in any case, so diminishes the world that it mocks one's humanity." While individuals need to be able to obtain and perform a job, and nations need economies, such objectives make for a rather dreary and uninspiring education.

Global competitiveness is not the only game in East Asia. In half the chapters, a second story, which I'm calling *scientific literacy for civic participation*, appears alongside it. This story is captured best by So, Wan, and Chu's opening of their chapter on science education in Hong Kong with the recent collapse of a primary school "green roof." The authors use this event as an opportunity to reflect on the risks inherent in technological advances and the need for children "to be equipped with the necessary scientific knowledge and skills starting from primary education to assess everyday and social issues, to have good reasoning for making sound judgement, and to voice their opinions in an informed and constructive manner." By equipping pupils with this knowledge and skill set, the authors hope to "prepare pupils to participate in public discourse in science-related issues."

This emphasis on civic participation, which appears also in the chapters on Taiwan and Korea, is reminiscent of one of the stories proposed by Postman, entitled the "American Experiment." However, whereas the ideal of scientific literacy for civic participation in Hong Kong, Taiwan, and Korea is presented as the relatively straightforward application of scientific knowledge and reasoning skills, in Postman's version, the idea of democracy itself is problematized. "The American Constitution is not a catechism, but a hypothesis," writes Postman, suggesting that the point of civic participation is to question the state and social order rather than consecrate them. For Postman, the American Experiment is about vigorous, continuous argument, about clashing with others over socially and politically charged issues. One wonders how such a story might play out in the more conformist East Asian societies – and in some of the more autocratic East Asian polities?

Alongside these two narratives, other possible stories are alluded to as the authors mention aims such as the "development of a healthy lifestyle" (Korea), "climate change and planetary ethics" (Singapore), a "zest for living" (Japan), "equal rights and chances to study science" (China), learning to think independently (Japan), and "enthusiasm in the potential of scientific inquiry" (Taiwan). But none of these themes are developed into a full-fledged story that might give meaning and depth to the study of science above and beyond becoming a scientifically literate worker and citizen.

By way of contrast, consider how Postman speaks about science and its study in his discussion of "Fallen Angel," one of five alternative narratives he proposes. According to the "Fallen Angel" story, we and our knowledge are fallible, but we have the capacity to correct our mistakes "provided we proceed without hubris, pride, or dogmatism":

> Taking this point of view, we may conclude that science is not physics, biology, or chemistry— is not even a "subject" — but a moral imperative drawn from a larger narrative whose purpose is to give perspective, balance, and humility to learning.

In my mind, such a vision has the potential to inspire the learning of science for its own sake, as a topic which is an end in itself, since it is intricately tied to our humanity and its perfection. I imagine that this is the sort of "existential purpose… to which an education in science ought to be legitimately subsumed" that Lee referred to in his critical discussion of topics that have not received sufficient attention in Singaporean science education. In my reading, all of the educational systems appearing in the book would benefit from such attention to the ends of science education. East Asia appears to have adequately addressed the technical questions of science education (at least relative to the competition); such lofty standing at the top of the global league tables presents an excellent opportunity to engage in metaphysical reflection.

References

Hanushek, E. A., & Woessmann, L. (2016). Knowledge capital, growth, and the east Asian miracle. *Science, 351*(6271), 344–345. https://doi.org/10.1126/science.aad7796.

Komatsu, H., & Rappleye, J. (2017). A new global policy regime founded on invalid statistics? Hanushek, Woessmann, PISA, and economic growth. *Comparative Education, 53*(2), 166–191. https://doi.org/10.1080/03050068.2017.1300008.

Postman, N. (1995). *The end of education: Redefining the value of school*. New York: Knopf.

Chapter 9
Some Comments on Primary Science from Outside the East Asian Region

Terry Russell

9.1 Introduction

Science is important and significant in people's lives. As the Royal Society affirms: *The scientific endeavour is one of the crowning achievements of humankind, contributing to our culture and civilisation and to improving our quality of life* (Royal Society 2012). It follows that so too must science education be vital. Yet in our current "Information Age," there is turbulence. This is not the productive "disruption" favored by the early adopting proponents of change so much as an erosion of reliability in the public exchange of knowledge. The spectrum of concern stretches from internet stories lacking provenance, without reliably identifiable sources or corroboration, to sources that promote deliberate disinformation. The authors of misleading web information may have an agenda with respect to contentious issues that are of enormous relevance to the public: the response to global warming, use of genetically modified crops, the implications of developing knowledge of the human genome, and so forth. Some websites may be the work of individuals motivated by personal gain who are posting web content purely as "clickbait," or lures whose main purpose is to generate advertising revenue from curious or gullible visitors. A sign of the times is that "post-truth" was admitted as word of the year to the Oxford dictionary in 2016 (https://en.oxforddictionaries.com/word-of-the-year/word-of-the-year-2016), defined as *relating to or denoting circumstances in which objective facts are less influential in shaping public opinion than appeals to emotion and personal belief.* Rhetorical language has always colored and enriched attempts at persuasive argument; appeals made "from the heart" are to be respected. Yet a statement by Massachusetts Institute of Technology (MIT) academic staff is evidence of alarm at the potential erosion of the status and role of science in knowledge

T. Russell (✉)
University of Liverpool, Liverpool, UK
e-mail: T.J.Russell@liverpool.ac.uk

Y.-J. Lee, J. Tan (eds.), *Primary Science Education in East Asia*, Contemporary Trends and Issues in Science Education 47,
https://doi.org/10.1007/978-3-319-97167-4_9

generation and public debate. Setting aside the political agenda within the wider MIT statement, it stands as a particularly emphatic definition of what science should be:

We uphold the principles of the scientific method, of fact- and reason-based objective inquiry. Science is not a special interest; it is not optional. Science is a foundational ingredient in how we as a society analyze, understand, and solve the most difficult challenges that we face (http://www.mitvalues.org/).

In offering my thoughts on primary science education in East Asia, special consideration of the fundamental "epistemic" aspect of science is therefore opportune. When and how do we enable pupils to think like scientists when faced with competing knowledge claims? The development of this epistemic dimension of science education – how we encourage children to learn to reason and weigh the evidence for their beliefs – warrants particular attention. This issue will be touched on at various points in the discussion that follows.

The East Asian chapter authors have each met a daunting challenge in producing their texts. To present a coherent picture of the state of primary science education within each of the six jurisdictions in East Asia represented is a formidable task. I am honored to have been invited to offer comments from outside the region, but I do not claim my views to be representative of the UK, or even of England. Reading these contributions brings home the complexity and diversity of the stories told by their authors within each chapter. My initial assumption about attempting an overview commentary has moderated toward a more realistic goal. Rather than attempt a form of synopsis or summary overview, I offer some overarching observations relating to issues in primary science that have been important in my own work and resonate with some current concerns and interests. I have always found interactions with international colleagues to have been a source of mutual insight, inviting further reflection as to how primary science could or should develop. I trust this book will serve to initiate further debate, stimulated by the alternative perspectives presented.

Common to all the chapter offerings is the acknowledgement that, for several reasons, science makes an important contribution to every child's education. As one of the STEM (Science, Technology, Engineering, and Mathematics) subjects, science contributes to the foundational knowledge that is essential to students who will join the part of the workforce that contributes in a major way to each state's economic development. Technological progress needs a stream of students pursuing career involvement in mainstream as well as "blue sky" science development and applications. All chapter authors recognize science education as an investment in their nations' futures. For those students who pursue careers other than in STEM subjects, science literacy is needed for informed engagement, whether at the level of participation in public debate or, in adulthood, of acceptance of payment of taxes to finance political priorities in scientific research and development. Science knowledge contributes to issues that impact in some way on every citizen. That science ranges from fundamental research about the material world through to understanding the place of humans and the planet we inhabit in the wider universe. And of course, every nation state looks at the performance of its own and its neighbor's students and asks the question, “How are our students performing? Can we do any more to support their progress?” It happens that the preparation of this book follows publication of ongoing analyses of the more recent TIMSS and PISA international surveys, both of which confirm the excellent performance in science of students in the East Asian region. This allows me to begin with an optimistic outlook about what has already been achieved. Beyond that positive sense, the program also provide useful sets of reference points which help to set the scene, though not with any assumption that international surveys offer the ultimate yardstick for measuring success. My assumption is that by looking behind the rankings and examining the detail, every participating system that is willing to scrutinize the results will be able to identify issues that warrant attention or invite strategies for improvement. There are many positive examples of positive reactions, as, for example, Korea’s response to the 2011 TIMSS survey that revealed high achievement levels but lower attitudes toward, and confidence in, science. As a consequence, “enhancing curiosity and interest towards science has been made a priority in the goals of the new 2015 national science curriculum.” Several other contributors make similar points.

9.2 Primary Science and International Monitoring Programs

The two international survey systems allow comparisons across participating countries, though participants are not the same in the two programs. The Trends in International Mathematics and Science Study (TIMSS) has conducted surveys every 4 years, 1995–2015; for some participants, a review over 20 years is possible for grades 4 and 8 (students around 10 years and 13 years of age). More than half a million students were sampled by TIMSS from 57 countries in 2015. The other survey, the Programme for International Student Assessment (PISA), has run surveys every 3 years since 2000, the sixth being in 2015 for reporting in 2016–2017. More than half a million 15-year-olds from 72 Organisation for Economic Cooperation and Development (OECD) countries were sampled and responded to a 2-hour test with a special focus on science. All of these PISA students would have been shaped by their primary school science experience in the 5–6 years through to 2004–2010, though by the time of the survey, many were coming toward the end of their formal education. All contributions in this book refer, to some extent, to these surveys. Overall, results are exceptionally positive for the East Asian region, occasionally outstandingly so. The comments on survey outcomes that follow will touch very briefly on only a few broad issues; it is to be expected that each participating authority will scrutinize the original within-country data in order to draw conclusions about issues and their trends over time. Ranking differences between countries and tracking changes in position over years are fascinating but must be treated with caution as changes in relative position are not always significant. Furthermore, average scores do not tell us what we need to know about underlying factors, so policy decisions need to be informed by other sources of evidence about teaching and learning outcomes. It is also fair to mention that the surveys have their critics in respect of their methodologies and there are those who point to the exaggerated impact they sometimes have on policy and resource allocation, but they certainly serve to open up the debate. For this reason, some issues arising are useful in offering a broad starting orientation to what follows.

9.2.1 Equity

It is well-established that a family's social class or overall economic health is a major determining factor in a child's educational progress. National averages are likely to mask differences. The educational vision is that each student's performance outcomes should reflect their talents, aptitudes, capabilities, and effort, rather than their family and social circumstances. "Resilient" students are defined as those who overcome disadvantageous odds by overcoming social handicap to perform at high levels. Economic disparities between different regions within a nation exert an influence on the overall mean school performance of regional populations. For example, in the UK, performance in England, Scotland, Wales, and Northern Ireland needs to be scrutinized separately, as there are occasional significant regional differences. Even within England, the relatively affluent Southeast outperforms the relatively poorer North of the country. Similar discrepancies are likely in the nation states reported in this book, perhaps in a more extreme form in some instances, less so in others. For example, in China (represented by four provinces in PISA 2015, Shanghai, Beijing, Jiangsu, and Guangdong), basic geographical factors in a province such as Guizhou in the Southwest can be understood to impose different starting conditions as compared with, for example, the more developed world-facing Shanghai. Investment is required to achieve equity of opportunity and access to resources. In the world's most populous country, rapid urbanization has inevitably created disparities between rural and urban dwellers, (OECD 2016a, b). In their chapter, Hu and Shou refer to a successful strategy employed in China (and no doubt elsewhere) of "mandatory administration" designed to level up rural schools with their urban counterparts. Each system must look at its local circumstances and plan its interventions accordingly in order to offer pupils a "level playing field." In England, one priority is to reduce the "tail" of lower performers at age 15 that we must presume began to show toward the end of the primary years. The PISA performance gap in England between the highest and lowest achieving pupils in science is unacceptable, being greater (in a relatively high science performing country) than in many other OECD countries. Similarly, while Singapore is the highest achieving country for TIMSS Year 5 science (and maths, incidentally), it has a relatively wide spread of achievement. Interestingly, despite the long "tail," students in England show a greater resilience to socioeconomic disadvantage than is the case in most other nations. Each jurisdiction has the opportunity – perhaps we should suggest *obligation* – to scrutinize PISA outcomes and use them formatively to inform local policy initiatives. To have an effect, such interventions need to be implemented from the earliest years of education.

9.2.2 Gender

Gender differences tend to be pervasive and entrenched in science, with boys more likely to be the top performers, as well as more likely to be among the poorest performers. There are differences in the kind of science favored as career choices by girls and boys. The fact that in Finland (an overall high PISA performing country) girls are more likely than boys to be top performers is a salutary indication of what can be achieved as well as a warning against complacency. The TIMSS 2015 performance outcomes suggest that the East Asian systems might find room for improvement in gender-related outcomes. Every system must be vigilant in guarding against the systemic and institutional gender stereotyping of what it is to be and think like a scientist. Such attitudes can be so pervasive as to be invisible, totally out of conscious awareness. In some subcultures, sex role stereotyping might even be considered appropriate and the correct way to behave. For example, at the time of writing, a children's toy distributor in the UK faces criticism for marketing dressing up sets for children: "princess" costumes for girls and "doctor" outfits for boys. Such typecasting is no doubt repeated in different forms throughout the world and may have an impact on career choices. Sex-role stereotyping may be amplified by the commercial interests of the fashion, media, sports, and entertainment industries. It is for science educators to be watchful – not to restrict choices for girls and boys but to widen their choice of career possibilities. Female students should not face any disincentive to pursue a science-related career, whether it is an ambition to pursue medical or any other forms of training.

9.2.3 Multiple Outcomes

Overall, PISA science performance is reported to have "flat lined," 2006 to 2015. The OECD summary of policy recommendations does not single out particular interventions for individual countries. From the generalizations in the data, it would be for each contributor to this book to decide which items on the list are relevant to their local situations: increasing support for disadvantaged students and schools, delaying selection processes applied to students, investment in the development of teachers, and challenging gender stereotypes. All solutions to problems and strategies for improvement must, of course, be viable at the local level. One attention-grabbing graphic in the mass of data is reproduced (with acknowledgment to OECD/PISA; see http://www.oecd-ilibrary.org/education/pisa-2015-results-volume-i_9789264266490-en), as Fig. 9.1, "multiple outcomes." This shows three outcomes as overlapping sets: students' mastery of the most advanced scientific problems, career aspirations in the direction of science-related employment, and demonstration of the capability to think like scientists (the epistemic dimension). Figure 9.1 is a reminder that a student's high science test (or examination) score is not an end in itself. The capability to pass an examination has to be

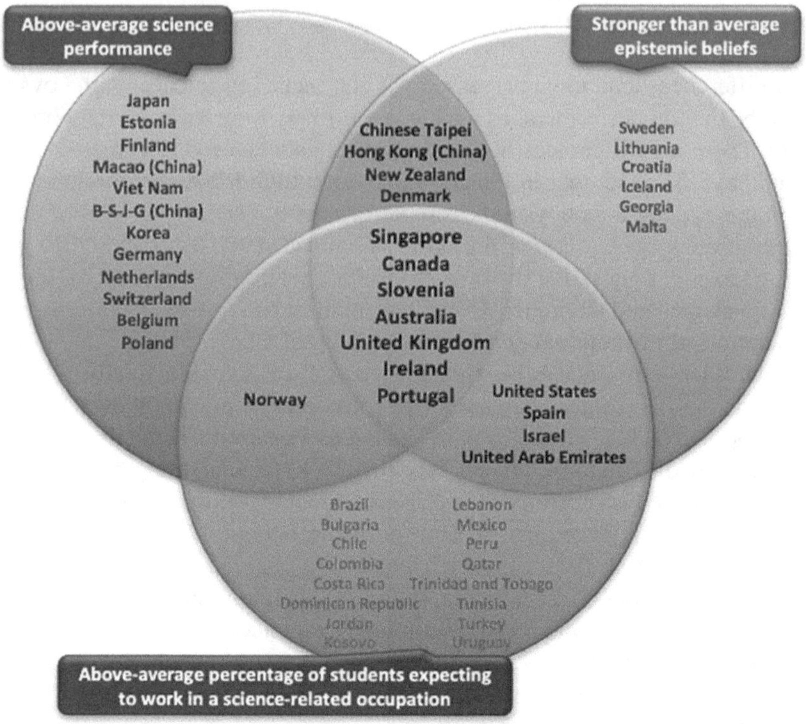

Fig. 9.1 Multiple outcomes. (From http://www.oecd-ilibrary.org/education/pisa-2015-results-volume-i_9789264266490-en)

translated into practical application. We also need to enthuse our pupils in their involvement in science, encourage them to use their capabilities, and make opportunities known to them. Even from their earliest school years in primary education, we can encourage pupils' use of evidence-based analytical thinking in all aspects of their lives and across the curriculum, as well as open up awareness to science career prospects. Figure 9.1 is interesting in setting science education in this wider context, and once again, East Asia is well represented in the central overlap of the three sets, with Singapore especially so.

9.3 The Nature of Primary Science

The basic economic relevance of a strong education in science together with assumptions about the need for a scientifically literate citizenship is almost universally accepted. The more difficult question is to decide in what form and manner primary science education can play a supporting role. Those structures and functions are not self-evident; they require empirical feedback about what works in practice and value decisions to be made about the nature of the primary science experience that is to be prioritized. It might be easier to decide what primary science should *not* be: it is clearly not to offer diluted versions of particle physics, biochemistry, or whatever, to young children. So there is a need to reflect on the question, *What is primary science?*, to think about how it is defined and to consider what are its constituents. Readers of the contributions to this volume, as well as consulting relevant books and journals from around the world, will rapidly appreciate that the definition is complex and multidimensional – and often the subject of intense debate. Furthermore, the definition is not static; it has shifted over time and, at an accelerated pace in recent years, influenced by the ubiquitous 24-hour, 7 days a week access to information. The consequence is that there will be differences in curricula and practice, implicit or explicit, across national borders. As well as cultural variations, these differences will reflect alternative emphases, influenced by how science is seen in the context of the wider curriculum and longer-term educational objectives.

If primary science is to be something other than "watered down grown-up science," then what should it look like? What do the specialists offer in place of such an assumption? The age range that is our concern, broadly 5–6 to 11–12 years of age, is universally acknowledged as a period during which an exceptional rate of human biological change occurs, a spurt in growth that educators must work with, not against. The age range encompasses the period from the time children are ready to step out from the intimate care of the home toward greater independence, through to early adolescence and the threshold of maturity. The focal age range that is the subject of these chapters encompasses a period of very particular development and progression. It is in this context that the question arises, *How does primary science deal with progression?* Rather than think of development between primary, secondary, and higher science education as step changes, what course can educationalists plot that ensures continuity in students' learning experience? My background as a developmental psychologist has stimulated my interest in the exploration of curricular continuity in teaching and learning and progression in understanding. These have been constant themes in my research and development activities.

9.4 Evolution of the Constituents of Primary Science

Science is a relative latecomer to the primary curriculum and consensus as to the constituents of the subject may not yet be established. Lack of unanimity is in part due to a considerable widening recently in the range of capabilities considered necessary to be included. Various of those aspects might be prioritized in any particular system. In considering science as a core subject alongside literacy and mathematics, it is helpful to trace the history of its development. This developmental history will, of course, vary from country to country.

In the UK, science in the primary curriculum was gradually taken up from as recently as the mid-twentieth century. The well-established secondary science curriculum had been considered, quite correctly, to be inaccessible and inappropriate to extend downward to the 5–11 age range. It took a fundamental shift in educational thinking, stimulated in part by the emergence of cognitive developmental psychology as a discipline, to reveal exciting new educational possibilities for young children's learning. Piaget and Inhelder's (1969) research into the development of thinking and intelligence was primarily a form of empirically informed philosophy. They listened to and observed the manner in which young children solved real-world problems concerned with number, mass, area, volume, density, speed, movement, and so forth. Their primary interest was in epistemology, human developmental knowledge construction, and they conceived the emergence of intelligence as a form of biological adaptation by the human organism. It was for educators to recognize the relevance of "logico-mathematical" thinking and its application to the teaching of mathematics and science in the primary curriculum (Isaacs 1962). From early in the introduction of primary science in the UK, there has been a back and forth debate: should we prioritize the "what" of science (science concepts) or the "how" (science processes)? The proponents of emphasizing the "what" side wished to lay the foundations for the big ideas that have shaped modern culture (evolution, the solar system, electricity, chemistry, energy, etc.), while the "how" advocates favored introducing children to science processes (the ways of experimental inquiry practiced by scientists). In England, this fluctuating debate has matured into the current national curriculum with pupils expected to be taught the "knowledge, methods, processes and uses of science" (DfE 2015). The programs of study describe a sequence of knowledge and concepts, with "working scientifically" expected to be integral to all work with science subject matter. Pupils must be helped to develop *understanding of the nature, processes, and methods of science for each year group*. We need to question what exactly the "nature, processes, and methods" of science are. The UK wording is perhaps deliberately ambiguous as previous incarnations of the curriculum having greater specificity have led to teachers protesting, "Overload!" There is help at hand by looking further afield to locate attempts at being more specific about, for example, what the nature of science entails. What is clear is that the original epistemological concerns remain an important item on the agenda.

The majority of UK primary teachers' apprehensions with respect to teaching science are well known, and so it is not surprising to hear a similar situation confirmed in East Asia. The exploration of what is expected of classroom teachers and whether these factors are manifest in this book's chapters will need to get more complicated before the mist clears. Science education researchers have attempted to elucidate matters by being more specific about exactly what the "nature of science" (NOS) entails in the way of pupils' understanding. Researchers such as Lederman et al. (2002) have contributed to the debate about the essential philosophical and epistemological components of science and refer to the notion as multifaceted, complex, tentative, and dynamic. Though consensus is elusive, following Lederman (op cit) the set of understandings that constitute NOS can be restated as seven ideas:

1. (*Empirical*) Science is empirical in nature, built on the evidence of observation and measurement of the physical world.
2. (*Theory/laws*) Scientific theory and scientific laws are different and need to be clearly distinguished.
3. (*Creative*) Science is interpretative, requiring sensemaking and model building; creativity and imagination play a central role in the creation of scientific knowledge.
4. (*Theory-laden*) Scientific knowledge is theory-laden in nature (and the distinction between observation and inference must be learned).
5. (*Social*) Scientific knowledge is socially and culturally embedded and the product of exchanges of understandings.
6. (*Multiple methods*) There is no single scientific method but rather a range of possibilities.
7. (*Provisional*) Science knowledge must be regarded as tentative and provisional in nature rather than certain and never-changing.

The list above is intended to apply to all science, without considering possible limitations in the case of young learners. Nonetheless, six of the criteria can be translated into activities that are realistically accessible to the primary age range, the exception being the second: "making the distinction between theories and laws." No harm will be done by holding off on this criterion until secondary education.

Another set of expectations to contribute to the debate derive from the USA's Framework for K-12 Science Education (National Research Council 2012), which refers to eight scientific (and engineering) practices. Although there is much in common with the UK position, the US document sets out the components in a more explicit fashion, which is useful for this discussion. (It also carries the danger of over-specification and a mechanical list-ticking approach to teaching.) This Framework document describes the frustration with "inquiry science" in the USA as driving the desire for a new specification. Dissatisfaction was attributed in part to a tendency for the students' investigations to lapse into programmed teaching sequences that were required (in teachers' application of the approach) to result in the correct textbook outcomes. Restated, rather than an inquiry approach that promoted thinking as scientists, pupils were being stage-managed by teachers toward predetermined outcomes. Every one of the East Asia chapters confirms that inquiry (or "enquiry" in UK English) approaches are favored and have been widely adopted, though not always with satisfactory outcomes. For example, the contributions from China and Japan recognize the challenges as well as the possibilities inquiry methods pose for teachers. In Taiwan, expert teachers are reported to have been deployed and inquiry-based competitions organized to stimulate and enhance the level of professional expertise required. The Singaporean chapter refers to the dual attraction of inquiry methods in offering opportunities to stimulate students' procedural *and* conceptual understanding. The interesting but disappointing outcome is the observation that, in one reported study, serving teachers felt the pressure of focusing on knowledge acquisition rather than procedural practices, bowing to perceived pressure to prepare students for content recall in examinations. And of course, all around the world, the fact is that the practical competencies that are integral to inquiry approaches are more difficult and expensive to assess in summative tests and examinations and so tend to be missing from the assessment agenda. If not assessed, practical work is always in danger of slipping down teachers' order of priority.

Inquiry science, which had a strong advocacy for so long, is unlikely to be displaced from primary classrooms and nor should it be. What England's "working scientifically" and the US Framework documents hope to achieve is a better balance, the avoidance of overemphasizing practical inquiries to open up curricular space for a more nuanced set of practices that represent a more realistic spectrum of scientific behaviors. The eight US Framework practices are defined as follows:

1. Asking questions (for science) and defining problems (for engineering)
2. Developing and using models
3. Planning and carrying out investigations
4. Analyzing and interpreting data
5. Using mathematics and computational thinking
6. Constructing explanations (for science) and designing solutions (for engineering)
7. Engaging in argument from evidence
8. Obtaining, evaluating, and communicating information

In addition to the eight scientific practices, the document identifies seven crosscutting themes (patterns, cause and effect, etc.) and "disciplinary core ideas," otherwise known as subject matter knowledge. As the US Framework applies to 12 years of science education, its interpretation and application demand a route through that reflects progression in accessibility and comprehension, year by year, for primary-age pupils. The document sketches how competencies might progress across grades but is honest in admitting *there is very little research evidence as yet on the developmental trajectory of each of these practices*. This admission offers a set of research agendas for science educators internationally.

Given these revisions in conceptions of science education, it is of interest to monitor signs of similar impacts of any of the seven NOS criteria or eight framework practices in national curricula in the East Asia region. In the chapters in this book, the sense is of a scenario similar to the position in the science curriculum in England: they can *all* be located, in some manner, but often in implicit rather than explicit form. "Working scientifically" within the national curriculum in England is so broadly defined as to include all the elements discussed. This kind of fuzzy boundary around the definition of what science is (or should be) gives license to research and development that will need to explore what is effective for teachers and students in pursuit of a rounded science education, without destabilizing mainstream efforts where there is confidence in the existing agenda. There is also the danger to be avoided of mechanical approaches that treat each element as a discrete subset for directed teaching, rather than overlapping and interacting qualities. Referring back to Fig. 9.1, it seems very likely that students' capability to think like scientists (the epistemic dimension) will benefit from gaining in prominence in coming years, in policy decisions, curriculum specificity, and teachers' practices. Examination of the first six US Framework practices does not suggest anything that would be unfamiliar to any of this book's authors. There might be quibbles, as, for instance, a personal view that expectations of "developing and using models" need to be deconstructed from a development perspective lest expectations become unrealistic. More static rather than dynamic "images in the mind" are often more appropriate for the primary age group, in the form of iconic mental representations. (This discussion is elaborated in greater detail in Russell and McGuigan 2016a.) The more innovative additions in the list of practices are the seventh (*engaging in argument from evidence*) coupled with the eighth (obtaining, evaluating, and communicating information). These are novel in that they expand the repertoire of classroom possibilities for primary science to include exchanging ideas, reading, writing, and discussion, the latter more formally expressed as "argumentation," where the emphasis is on ideas discussed as tentative claims that require the support of evidence. This stance is a direct challenge to approaches in which science education's goal is *primarily* arriving at a body of factual knowledge.

It is fair that teachers should expect practical guidance rather than definitional conundrums, and in this sense, there is some comfort to be gained from the earlier US consultation, "Taking science to school" (NRC 2007). That document sets out a simpler formulation in the form of just four strands of capability for the phase of kindergarten to Grade 8 and helped set the scene for the later US Framework:

1. *Procedural:* Generate and evaluate scientific evidence and explanations.
2. *Conceptual:* Know, use, and interpret scientific explanations of the natural world.
3. *Communication:* Participate productively in scientific practices and discourse.
4. *Epistemic:* Understand the nature and development of scientific practices and discourse.

The *conceptual* and *procedural* domains are familiar from the earliest days of primary science but are currently emphasized in the UK and elsewhere as operating in partnership, rather than competing for primacy. The addition of *communication* has the virtue of connecting science to the wider primary curriculum, in particular literacy. It also acknowledges the nature of scientific progress through debate, journals, and books. The recognition of the *epistemic* aspect as relevant even within the earliest years of education is important.

We can go further toward establishing coherence across all the sets of components introduced in the discussion above if, instead of referring to transactions around *science facts*, the pedagogical standpoint is to assume an attitude of social discourse around diverse *tentative ideas*. This shift from notions of science "facts" to "ideas" – defined as requiring evidence to support them – has profound significance for primary classroom practice. The switch is in treating pupils' (and teachers') own science ideas as *provisional* and needing to be backed by *evidence*. Later, in secondary education, these tentative ideas can be referred to as *claims* and be more fully informed by science content knowledge. Claims require *evidence* for their acceptance or rejection; they can be challenged. This shift in pedagogy is radical in laying the foundation for the scientific mode of discourse, that of *scientific argumentation*, where claims have to be backed by evidence and may need warrants in defense against counterclaims. Constructivist research has long acknowledged the legitimacy of science educators benefiting from treating children's ideas seriously, so a shift in the direction of considering those ideas as if they are claims and requesting evidence from pupils to back them up is an easy and simple step, pedagogically. Even without the vocabulary of "argumentation" and "claims," the style of discourse implied can become rapidly and intuitively manageable to teachers and accessible to children. (How this can work in practice is discussed in Russell and McGuigan 2016b.) The important modification is in teachers' and children's attitude toward science ideas *as though* they are claims. This means treating all expressions of science belief as provisional in nature, always requiring supporting evidence – in other words, behaving in much the same manner as adult, professional scientists.

Productive dialogic interactions between teachers and students and between students themselves are increasingly recognized by science education researchers as important in establishing deep understanding. Christine Howe's research (2014) confirms the progress children can make in small group dialogue, even in the absence of teachers. While this orientation focusing on dialogic and argumentation has been evident in science education journal reports for several years, the approach probably remains more tentative and at the edge of thinking as far as classroom implementation is concerned. The chapter contributed from China echoes this same shift in reporting the desire to "transform from prioritizing knowledge-intensive learning to dialogic modes; building and emphasizing the interaction between teachers and students during teaching." As apparent in the various country reports, in general, education systems' curricula do not undergo revision very frequently, not least because the knock on effects of change can be expensive in training. This is even more so in contexts where textbooks underpin teaching practices and require replacement and rewriting. The extent to which epistemic practices generally and argumentation in particular are referenced in this volume is not extensive. Such awareness is explicit in the contributions from Hong Kong, Singapore, and Taiwan, and small-scale exploration of argumentation is reported from Singapore. This is not to suggest a lack of awareness or relevant activities in these or the other remaining contributions, only that the topic appears not to be in the foreground of current practice. Managing a focused classroom argumentation session is demanding even for experienced professionally trained teachers – rather like chairing a meeting in which each participant may have brought their own agenda of preconceptions. Engaging pupils in productive argumentation is best thought of as a common way of working across a school:

Establishing confidence with argumentation is best seen as a whole-school, long-term, cross-curricular project. From their first entry into education, we must assume that every child has ideas to contribute and build a supportive climate to avoid apprehension and facilitate expression. (Russell and McGuigan 2016b, p. 30)

9.5 Emergent Science

Once science educators become sensitive to the concept of developmental progression, that awareness raises the question, *When should primary science education begin?* (Incidentally, we might also inquire when it is cost-effective to *cease* mandatory basic science education, given the need for an at least minimally science-literate citizenship and the need for a supply of students with sufficient science understanding to train as teachers of science to young children.) In one sense, young infants' science understanding of the world is independent of adult supervision: they explore and handle everything with which they come into contact. The familiar act of infants dropping objects from a high chair and craning to see where and how they have landed has been acknowledged as a universal phenomenon and perhaps the prototype of inquiries into the nature of gravity. More formal efforts have been made in recent years to establish an agenda for science learning in the early years, designated as "emergent science." This term is well-established in the UK but perhaps less so internationally, represented by the open access Journal of Emergent Science (JES) (ASE 2017) under the auspices of the UK's Association for Science Education. (Since its recent support to the ASE in publishing JES, the Primary Science Teaching Trust has – surprisingly – extended the age coverage to 0–11 years, within which "emergent" science articles still appear.) In contrast to this development, it seems that most of the contributions to this volume refer to science education as being introduced at around the age of 8. It may be that authors have adopted a particular, more restricted, definition of primary science; there are initiatives described in the 0–6, 3–6, or 3–8 age range that address the beginnings of science education. The kind of developmental perspective that I have argued for necessitates awareness of antecedents and subsequent developments.

For the youngest age group, science has to be viewed within the holistic early years phase, so the wider context needs to be understood. Education is compulsory and subject to regulation and quality control for all children in England between the ages of 5 and 16. (This varies slightly across the countries that make up the UK, but what follows will refer to England.) Since 2010, all 3- and 4-year-olds have been entitled to 15 h of free nursery education for 38 weeks of the year. The "settings" in which early years education takes place have various forms: state nursery schools, nursery classes, and reception classes, are offered within state primary schools. Outside the state sector, operated under voluntary or commercial management, are childminders, voluntary preschools, and privately run nurseries. These forms of early years education and childcare have multiplied since the introduction of the "Foundation Stage" in September 2000 and the Education Act 2002. The latter extended the national curriculum to include the Foundation Stage that sets out the regulatory framework for education from the age of 3 to the end of the reception year when children reach the age of 5, the year before compulsory education begins. The Early Years Foundation Stage (EYFS) came into force in 2008 and is a single framework for the provision of learning, development, and care for children in all registered early years settings between birth and the school year they turn 5. There is a statutory EYFS profile (EYFSP) of development and learning achievements at the end of the school year children reach 5 years. The framework emphasizes the needs of each individual child to have experience of positive relationships; curricular expectations include language, physical and social development, literacy, mathematics, arts, and science (DfE 2014). It has to be acknowledged that starting ages, curricular experience, and the assessment process have been and remain the subject of considerable debate in England. It is also the case that the diverse range of preschool settings for 0–5-year-olds is staffed by people with equally varied forms of training, with over one tenth untrained and suitably qualified practitioners being about 75% of the workforce.

In recent years, educators have recognized that, in a facilitating environment, young children's learning can be an active, self-directed activity (NRC 2007). Young minds do not wait for adults to dictate when they are ready to start making sense of their world. Seen within the broader process of human evolution, sensemaking is now recognized as a drive toward intellectual adaptation to the environment. Opening up the agenda of what constitutes science to include the epistemic dimension – what children believe and their awareness of the need for a justification for their beliefs – makes participation in scientific ways of behaving of the youngest age group so much more possible. Among those practitioners having a fragile sense of their own science background, a familiar strategy when uncertain how to respond to children's questions has been to turn the question back in the form, "Let's find out!" This would be a prelude to some form of practical inquiry, usually involving "fair testing," that would allow the child the opportunity to explore with support while incidentally "buying time" for the teacher to carry out some personal background research. This strategy can work, though an established science background makes managing children's science learning easier, through a greater awareness that pedagogical content knowledge (PCK) provides, of possibilities for moving their understanding forward. As the Hong Kong chapter puts it, background science helps "in responding to students" questions, preparing teaching materials or tools, clarifying scientific concepts, focusing student attentions on learning, and organizing scientific investigations in the classroom."

Many early years educators will emphasize the need for children to have the freedom to socialize in unstructured play situations. Imaginative play need not be thought of as incompatible with or even contrary to the values of early childhood practices. It is useful for children to be able to distinguish between the real and imaginary worlds. The important recognition comes with practitioners' appreciation that multimodal forms of expression – speaking, writing, drawing, modeling, miming, etc. – can be treated as offering examples of beliefs or "knowledge claims." With this stance, it follows that it is legitimate for teachers and carers to ask, "What makes you think that?" or "Why have you drawn this object in this way?" Young children may push away such implicit challenges to their beliefs, however gently expressed, but the habit of adults asking such questions can soon become established and children learn the style of what is expected of them (Russell and

McGuigan 2016c). In fact, children can readily develop the habit of openness about seeking justifications or evidence for why things are as they are by using the same style of questioning between themselves, as a form of peer review.

9.6 Assessment

Peer- and self-assessment are at the "low stakes" end of the assessment spectrum. The enormity of the topic of assessment is such that this chapter could dwell on the subject to the exclusion of all other issues. Every chapter contribution refers to test and examination regimes, in recognition of the reality that valid and reliable assessment makes an essential contribution to successful primary science teaching and learning. In some cases, reference is made to traditional national systems with long histories that color modern expectations. Assessment, tests, and examinations are touched on in their various forms, whether formative, summative, or diagnostic in purpose, as well as from the perspective of different stakeholders. The needs for monitoring the outcomes at the international, national, regional, or school level are very different as compared to the needs of individual pupils and their families. Mostly, the different purposes require different instruments, the meaning of the outcomes requiring technical understanding if apprehension and confusion are to be avoided. Technology can play an increasing role in recording and archiving performance for later retrieval. Individual digital portfolios promise archives of pupils' achievements and future learning goals in personalized form. Technology also offers the potential to narrow the gap between formative uses – those that inform teachers about pupils' progress and current needs – and the summative function of the same data when retrieved to inform decisions or transition points in a student's educational career. The pressure that "high-stakes" assessment imposes on participants is widely acknowledged, whether felt by system administrators, head teachers, class teachers, or pupils. Some of this volume's authors also acknowledge that such pressure can put assessment and examination systems under enormous strain that lead to unfortunate "overheating," distortions, and attempts to subvert the system.

If education systems wish to capture the fine-grain reality of pupils' performance, the professionalism of teachers has to be trusted. The reason for this is simple to understand: teachers are likely to be the only direct witnesses of pupils' performance. For example, in the UK, it is acknowledged that some, but not *all*, aspects of "working scientifically" can be assessed reliably and validly through the use of pencil-and-paper testing procedures. It is possible to ask pupils to write about how they would plan an investigation within a described scenario or use particular apparatus such as thermometers or newton meters within an inquiry. The question arises as to the extent a written assessment can be a valid and reliable predictor of practical skills, even accepting the extraneous demands on reading and writing. Another consideration is that pupils' primary science investigations are usually conducted collaboratively. Teacher assessment is the only realistic strategy for capturing valid information of this kind, with each student's contributions needing to be observed over several sessions as they adopt different roles within the team effort. A series of observations carried out by a teacher within the normal primary classroom ecology will offer the conditions for a valid assessment. Another important aspect of "working scientifically" is communication. Presenting one's own results and interrogating the presentations of others are best achieved through classroom discussion or argumentation sessions in which pupils contribute their ideas (or "claims") coupled with supporting evidence to back them up. Here again, pupils' "thinking scientifically" is best witnessed directly over time in different contexts. Not least, we must be mindful of nurturing students' *enjoyment* of working and thinking scientifically, factors that will contribute to their satisfaction with engagement in the subject and which will enhance the possibility of science being pursued as a career choice. These affective dimensions relating to a sense of well-being, motivation, resilience in the face of impediments, and determination to persevere are very elusive to measurement by formal test procedures. They should not be assumed to be on a par with other measures of science performance in the TIMSS and PISA results for East Asian countries, where pupils' motivation and confidence tend to be less well developed.

Education systems need the feedback of summative data in order to monitor effects of whatever intervention their policies have implemented. Unfortunately, to teachers, the tests used to monitor systemic outputs are often seen as a threat or an incursion on what they regard as their core professional role of teaching children. The solution is to ensure that teachers feel that assessments, tests, and examinations are aspects of education that they are involved in, tasks done *with* them rather than imposed or done *to* them. To arrive at this understanding, teachers need to be confident in gathering and interpreting assessment information, both qualitative and quantitative. They need to understand the technical and measurement possibilities as well as the limitations of testing and the data it generates. This level of professionalism requires support and guidance, both in initial teacher training and in continuing professional development as systems develop, change, and mature. Only then can we hold realistic hopes of implementing teachers' own formative assessment practices and pupil self-assessment. This trend toward "more of formative, performance-based assessment to understand and enhance students' learning" is mentioned in the Korean chapter, while in China, it is reported that "ways of assessment are getting more diverse."

9.7 Summary and Conclusions

On the evidence of the various chapter contributions to this book, coupled with the outcomes of international monitoring programs, primary science in East Asia appears to be in extremely good health. Equally true is the fact that the world does not stand still. Economies boom and bust, financial circumstances influence political decision-making, and technology develops in unanticipated directions.

Complacency is dangerous, and all outcome measures need to be treated as formative data through which to scrutinize systems and see what needs attention or can be improved. The issues to which this contribution has drawn attention are discussed not as criticisms but as reflections on possibilities for yet further development and improvement. Nonetheless, congratulations are in order and well deserved. While some important issues have been aired, several other extremely important topics have not been addressed at all: teacher education (whether it should be specialist or generalist), professional development strategies, curriculum content coverage and progression, textbook development, and use of IT resources, CD-ROMs, blogs, video training, websites, and so forth. There is also the role played by science associations and journals that some authors touch upon. In the UK, the Association for Science Education (ASE) plays a key role in offering in-service education for teachers (INSET courses) as well as supporting and publishing a range of journals. The most important of these for the primary range is *Primary Science*, published five times a year and aiming to support effective practice in science education focused at the primary school level through transition to early secondary education. The highlight each year is the ASE Annual Conference that circulates between various university campuses in England. Also important to UK science educators is the European Science Education Research Association (ESERA) which holds biennial conferences and offers particular support to PhD candidates. Both of these conferences attract international participants and offer very convenient venues for meeting and continuing the discussion of the topics stimulated by this book.

References

ASE. (2017). No boundaries, no barriers. *Journal of Emergent Science*, Special edition, Issue 12, Winter 2016/17. Available for download from http://www.ase.org.uk/journals/journal-of-emergent-science/

DfE. (2014). *Statutory framework for the early years foundation stage setting the standards for learning, development and care for children from birth to five.* Published March 2014 Effective September 2014 Reference: DFE-00337-2014 https://www.gov.uk/government/uploads/system/uploads/attachment_data/file/335504/EYFS_framework_from_1_September_2014_with_clarification_note.pdf

DfE. (2015). *Statutory guidance. National curriculum in England: science programmes of study department for education.* Updated 6 May 2015 © Crown copyright https://www.gov.uk/government/publications/national-curriculum-in-england-science-programmes-of-study/national-curriculum-in-england-science-programmes-of-study

Howe, C. (2014). Optimising small group discourse in classrooms: effective practices and theoretical constraints. *International Journal of Educational Research., 63*, 107–115.

Isaacs, N. (1962). The case for bringing science into the primary school. In W. H. Perkins (Ed.), *The place of science in primary education*. London: British Association for the Advancement of Science.

Lederman, N. G., Abd-El-Khalick, F., Bell, R. L., & Schwartz, R. S. (2002). Views of nature of science questionnaire: toward valid and meaningful assessment of learners' conceptions of nature of science. *Journal of Research in Science Teaching, 39*(6), 497–521.

National Research Council. (2007). *Taking science to school: Learning and teaching science in Grades K-8.* Committee on science learning, Kindergarten through eighth Grade (Duschl, R. A., Schweingruber, H. A., & Shouse, A. W., Eds.). Board on Science Education, Center for Education. Division of Behavioral and Social Sciences and Education. Washington, DC: The National Academies Press.

National Research Council. (2012). *A framework for K-12 science education: Practices, crosscutting concepts, and core ideas*. Committee on a Conceptual Framework for New K-12 Science Education Standards. Board on Science Education, Division of Behavioral and Social Sciences and Education. Washington, DC: The National Academies Press. PDF available at http://nap.edu/13165

OECD. (2016a). *PISA 2015. Results in focus.* https://www.oecd.org/pisa/pisa-2015-results-in-focus.pdf

OECD. (2016b). *Education in China. A snapshot.* http://www.oecd.org/china/Education-in-China-a-snapshot.pdf

Piaget, J., & Inhelder, B. (1969). *The psychology of the child.* New York: Basic Books.

Royal Society. (2012). *The royal society strategic plan, 2012–2017.* June 2012 DES2480 https://royalsociety.org/~/media/Royal_Society_Content/about-us/governance/Strategic-Plan.pdf?la=en-GB

Russell, T., & McGuigan, L. (2016a). *Exploring science with young children. A developmental approach.* London: Sage.

Russell, T. and McGuigan, L. (2016b). *Developing 'argumentation' with the 4–11 age range.* Primary Science No. 144 Sept./Oct. 2016.

Russell, T., & McGuigan, L. (2016c). An updated perspective on emergent science. *Early Child Development and Care, 187*(2, 2017), 284–297.

Chapter 10
Scientific Literacy in East Asia: Shifting Toward an Inquiry-Informed Learning Perspective

Umesh Ramnarain

The new social and economic demand for higher-order intellectual skills which has accompanied the global information technology revolution has impacted on school curriculum reform throughout the world. This is certainly the case in the East Asian countries that fall within the purview of this book on primary school science. National growth and competitiveness is dependent on continuous technological improvement and innovation, driven by a workforce that needs to be adaptable in its thinking and operate with greater autonomy. Whereas skills in set routines were valued as attributes in the past, today each worker is expected to think critically, solve abstract problems, and generate new ideas for improvement (United States Department of Commerce 1999). This means that people should acquire the capacity to constantly redefine the necessary skills for a given task (Castells 2000). These new demands from the workplace and the technological advancements of the world in which we live have served to stimulate much change in national curricula throughout the world. All six East Asian countries have a knowledge economy whose sustainability is driven by a highly skilled workforce in the fields of science and technology. Globalization of the economies and rapid advances in information technology mean that all countries must remain competitive for global market share.

In all six East Asian countries, there is a reconceptualization of scientific literacy from a curriculum dominated by content specification to one where the goals of scientific literacy are more broadly defined. According to Osborne (2000) and Hodson (2003), scientific literacy can be perceived in four different ways including (1) cultural, developing the capacity to read about and understand issues relating to science and technology in the media; (2) utilitarian, having the knowledge, skills, and attitudes that are essential for a career as scientist, engineer, or technician; (3) democratic, broadening knowledge and understanding of science to include the

U. Ramnarain (✉)
University of Johannesburg, Johannesburg, South Africa
e-mail: uramnarain@uj.ac.za

Y.-J. Lee, J. Tan (eds.), *Primary Science Education in East Asia*, Contemporary Trends and Issues in Science Education 47,
https://doi.org/10.1007/978-3-319-97167-4_10

interface between science, technology, and society; and (4) economic, formulating knowledge and skills that are essential to the economic growth and effective competition within the global market place. Curriculum reform initiatives in all six countries represent these aspects of scientific literacy. There is a definite paradigm shift from the traditional teacher-dominated to a learner-centered approach. Science is depicted as a human activity, and there is a strong emphasis on engaging the learners in the processes of sciences. This commentary will seek to extract defining principles in the curriculum reform for each country as enunciated the chapter authors. In many cases, the intentions of curricula are laudable and educationally sound; however, all too often the attention and energies of policy makers are focused on the "what" of desired educational change, neglecting the "how" (Rogan and Aldous 2005). In any context, the success curriculum reform being implemented at the micro level of the classroom is contingent upon teacher preparedness. Concomitant, this commentary, seeks to unpack readiness of teachers in giving expression to curriculum reform efforts. With regard to the above two aspects, a synopsis is provided for each country based on the contributions of the authors. Where obvious opportunities exist, comparisons are made across the countries so that defining principles may be expanded upon.

In the six East Asian countries, primary school science education has undergone waves of curriculum reform, with inquiry-based instruction being a defining feature in driving the goals of scientific literacy. This shift toward inquiry-based science education is in tandem with curricula reform efforts worldwide. Practical work in the school science curriculum, in particular inquiry-based learning, has received much attention in the curriculum reform initiatives which have taken place worldwide. One of the major changes advocated in this curriculum reform is a new conception of the role and form that practical work should assume. These developments in the East Asian countries that form the focus of this book mirror the worldwide reform trends in science education. In the United Kingdom, Attainment Target 1 for Science in the National Curriculum has apportioned much priority to scientific investigations (Department for Education and Employment 1999). In the United States, the American Association for the Advancement of Science (AAAS) and the National Research Council (NRC) endorse science curricula that actively engage learners using an inquiry-based approach. The National Science Education Standards developed by the NRC (1996), mandate:

> Students at all grade levels and in every domain of science should have the opportunity to use scientific inquiry and develop the ability to think and act in ways associated with inquiry, including asking questions, planning and conducting investigations, using appropriate tools and techniques to gather data, thinking critically and logically about relationships between evidence and explanations, constructing and analyzing alternative explanations, and communicating scientific arguments. (p. 105)

The AAAS (1993) in its Project 2061 recommends that before graduating from high school, students working individually or in teams should design and carry out at least one major investigation. In doing so, they should "frame the question, design the approach, estimate the time and costs involved, calibrate the instruments, conduct trial runs, write a report, and finally, respond to criticism" (AAAS 1993, p. 9).

In Singapore, the primary science curriculum is revised every 6 years. A renewed emphasis on the syntactical aspects of science means that inquiry-based science education has become a core feature of the curriculum. Although earlier versions of the curriculum had inquiry as a defining focus, the 2008 curriculum underlines a shift from a "hands-on" to a "minds-on" approach to inquiry. In this regard, the authors cite the Singaporean primary science curriculum where is it stated that:

> Teaching science as inquiry must therefore go beyond merely presenting the facts and the outcomes of scientific investigations. Students need to be shown how the products of scientific investigations were derived by scientists and be provided opportunities to: ask questions about knowledge and issues that relate to their daily lives, society and the environment; be actively engaged in the collection and use of evidence; formulate and communicate explanations based on scientific knowledge. (CPDD 2007, p. 11)

The predominance of the prevailing "hands-on" approach is certainly in concert research reported elsewhere where teachers typically involve students in routine practical activities where they slavishly follow teacher directions and procedures without much thought. Abrahams and Millar (2008) investigated the effectiveness of practical work by analyzing a sample of 25 typical science lessons involving practical work in secondary schools in the United Kingdom. They concluded that teachers predominantly focus on making students manipulate physical objects and equipment. There was little evidence of teachers supporting students in linking observations and experiences to conceptual science ideas. A similar finding was made in the United States. Students were not required to think about scientific explanations for their observations and results, with the result that science content ideas were only weakly connected to the activities. Science content, when present, was usually organized as a collection of discrete facts, definitions, and algorithms rather than a connected set of ideas (Roth et al. 2009). Chinn and Malhotra (2002) indicate that this practice reinforces an unscientific epistemology that promotes a belief that "science is a simple, algorithmic form of reasoning; as a result, students are likely to fail to learn the heuristics scientists use to reason under uncertainty" (p.213).

The reformed curriculum emphasis on the "minds-on" approach to inquiry mirrors developments in the United States where, in a recent publication of the National Research Council entitled "A Framework for K-12 Science Education," there is an emphasis on students experiencing inquiry-based practices and not merely learning about them (National Research Council 2012). Here, the term "practices" is used instead of "skills" to stress that engaging in inquiry requires the coordination of both knowledge and skills simultaneously. The following "practices" are identified: asking questions (for science) and defining problems (for engineering), developing and using models, planning and carrying out investigations, analyzing and interpreting data, using mathematics and computational thinking, constructing explanations (for science) and designing solutions (for engineering), engaging in argument from evidence, and obtaining, evaluating, and communicating information (National Research Council 2012, p. 42). This is now reaffirmed in the Next Generation Science Standards (NGSS; NGSS Lead States 2013) where scientific inquiry is now synonymous with a vision of scientific literacy that encompasses skills and knowledge related to scientific and engineering practices (Bartos and Lederman 2014).

Primary science in Japan also has a strong underpinning of inquiry learning. Here, the authors point out that the influence of inquiry is long-standing and is associated with the modernization movement of education, which was triggered by the Sputnik crisis in the United States in the 1960s. In the Japanese curriculum known as Course of Study for Elementary Schools (COSES), inquiry is located within the broader framework of problem-solving that encapsulates a series of steps of scientific inquiry that includes "contact with a natural phenomenon," "awareness/questions," "understanding the questions," "assumption/hypothesis," "experiment design," "carrying out the experiment," "obtaining results," "discussion," "conclusion," and "delivery" (MEXT 2006). A the local level, for example, in Kawashiri, the local curriculum called "Kashiwa Learning" conceives of inquiry learning as a means by which learning science at schools can be integrated with the daily experiences of learners. Indeed here, the one gets a sense of the intertwining nature of education and culture that is persistent with Asian educational settings.

In China, inquiry-based learning is expressed in the "Compendium of Curriculum Reform of Fundamental Education (1999)" through the notion that pupils regard themselves as investigators of science during instruction. In addition, pupils will make connections between what they learn in science lessons with the real world and experience or model some of the scientific habits/values of science (Poisson 2000). Further to this in the "Primary Science Curriculum Standards" of 2001, inquiry-based learning is expressed more explicitly where is regarded as one of the basic teaching aims as follows:

> Science inquiry is not the only goal of learning science, but also the content matter of science. The learning style advocated in science learning requires that students must actively engage in practical activities. Modeling how scientists investigate is one major goal of student learning in science.

Inquiry lessons would also appear to be timetabled, where from grade 3 there is one school-based science practical lesson to cultivate students' interest, inquiry abilities, and scientific literacy in science (Liu et al. 2012).

It is commonly regarded that any pedagogy is driven by assessment, and in China, this principle is given credence where assessment tasks on inquiry competency are described for teachers. Assessment is both formative and summative. Here, assessment tasks to detect students' ability in hypothesizing, observing, experimenting, measuring, giving scientific explanations, and appropriate communication can be accessed by teachers from science textbooks. Alternatively, teachers design these tasks by themselves based on guidelines from teacher guide books.

In China, the learning by doing (LBD) is a well-known experimental project aimed at the improvement of primary science reform. Similar to the practice in Japan, children's everyday experiences in their immediate environment is considered when teachers plan inquiry-based lessons. This is regarded is being critical in fostering links between scientific concepts and cultural awareness and knowledge. The LBD initiative has a goal that extends beyond impacting the science in the middle and high schools but is aimed at affecting the behavior and ways of thinking of the child by providing an enriching experience of science that influences the child both affectively and culturally.

As is the case in Singapore, inquiry-based learning is not being confined to the development of experimental, hands-on skills, but is also directed at the simultaneous coordination of both knowledge and skills. Here, it emerges that "Inquiry-based teaching is not just about processing or going through teaching resources, the main factor is for the students to participate in the identification of evidence, making judgments, reasoning logically, and thinking about how to interpret the data."

In Hong Kong, primary science is key learning area in an integrated subject called General Studies. The curriculum reforms are aligned with global trends in primary science education and suggest a shift from the tradition teacher-directed content-laden curriculum toward a curriculum that is learner-centered and inquiry-based. The authors indicate this as follows:

> The suggested curriculum development in GS is moving away from the content- focused approach and towards the learner-focused approach to enhance the inquiry and investigative skills of learners to help them construct knowledge, with an emphasis on the use of diverse resources and life-long learning opportunities.

In underlining the inquiry-based approach that is specified in the curriculum, the authors refer to the following except from the curriculum guide for General Studies for primary schools (Primary 1–Primary 6):

> students are active constructors of knowledge and the teacher is a facilitator of learning. ... students have to raise questions, find their own answers and look for the necessary information. They are engaged in identifying problems, collecting information and solving the problems they encounter. (pp. 75–76)

In addition to addressing inquiry through the teaching of curriculum content topics, learners are also encouraged to participate in science fairs and events such as workshops and competitions that are organized outside of school time.

As is the case with other five East Asian countries, in Hong Kong, there is also an imperative for teachers to use an inquiry-based approach to connect classroom science with experiences from the daily lives of learners.

The inquiry-based also dominates the primary science research agenda of Hong Kong. The authors point out that "The initial efforts of research on scientific inquiry focused on generating a theoretical model to conceptualize the process of scientific inquiry." Resulting from these efforts is a model proposed by So (2006) that identifies six components (i.e., ideas, design, interpretation, evaluation, understanding and application, and evaluation) of cognitive and practical processes that are regarded as necessary for doing scientific inquiry at the primary school.

Other research on inquiry has focused on readiness skills such as pupils' cognitive understanding and reasoning ability required (Lee and Ng 2004) and mathematical skills that are required for inquiry-based learning (So 2013). In addition, So and Ching (2011) designed a collaborative science learning environment to help teachers incorporate small-group learning with science inquiry. Limited research has been done on the classroom practice of inquiry-based instruction, with the authors only referring to a study by Yeung (2009) who analyzed the perception and implementation of inquiry in teaching of the teachers in two Hong Kong primary schools. It was learnt here that although teachers had a positive perception of inquiry, this approach was scaled down in implementation. The authors make a call for more government support to improve the culture of inquiry learning at the primary school.

It is also clear that information technology is playing an active role in advancing inquiry-based teaching and learning in a multimedia learning environment. An example of this that is cited by the authors is concept of a resource-based learning environment (RBLE) to develop pupils' understanding of family trees. They describe RBLE as comprised of three components, an audiovisual database for guided and coupled inquiry, an interactive interface for conceptualizing the nomenclature, and a tool for learners to construct their own family trees. The authors cite studies by Song and Kong (2014) and Song (2014) on the role of mobile devices. Here, it was found that when mobile learning was incorporated into scientific inquiry, pupils advanced their domain knowledge and developed their inquiry skills.

Inquiry-based learning is also prevalent in environment education that is referred to as the "youngest branch of research in primary science education in Hong Kong." For example, in the "Plastic Education Project," So et al. (2014) found that the inquiry-oriented strategies significantly improved the pupils' knowledge of the types of plastic waste and their beliefs about plastic waste. The authors encourage this interdisciplinary approach in science and environmental education and see it as having great potential to enhance the quality of primary science education.

In Taiwan, inquiry learning is regarded as a core characteristic of science education, and primary science teachers are expected to be resourceful in designing appropriate inquiry-based learning activities for their students. In the draft of the Curriculum Guideline of Science Learning Area of 12-year Compulsory Education document, one of the key dimensions of science learning is scientific inquiry abilities (thinking ability and problem-solving). In addition, the authors pointed out that according to the Grades 1–9 Curriculum Guidelines of Nature and Technology Learning Area, science education should be conducted using inquiry-oriented learning activities, which emphasize both hands-on and minds-on activities. Two of the main educational goals of the Science and Technology Learning Area at Grades 1–9 relate to inquiry-based learning (Chiu 2007, p. 309):

- To cultivate an interest in and a passion for science inquiry and the habits of active learning
- To acquire methods of inquiry and a basic competence in learning science and technology and to be able to apply one's learning to daily life

In contrast to Singapore where schools have well-resourced laboratories for inquiry learning, in Taiwan, most schools due to a tight budget have only basic equipment for practical work. Primary schools often rely on textbooks publishers for free basic laboratory equipment.

However, despite inquiry being a key curriculum goal, classroom practice remains teacher-centered with teachers being heavily reliant on textbooks. When practical work does take place, experimental tasks often embody a cookbook approach, where learners followed recipes for the execution of procedures handed down by teachers. Although these prescriptive exercises permit learners to learn how to follow "recipes" rather than develop the skill or understand the knowledge, the most crucial drawback of such exercises and other "cookbook" laboratory experiments is that learners do not know why they take certain steps over others. This approach is counter to learners acquiring an understanding of the nature of science.

It would appear that science fair projects provide a viable opportunity for learners to do open-inquiry activities where they have autonomy over designing and conducting their own inquiries. In fact, the authors allude to annual science fairs being deeply ingrained in Taiwanese school science with the first national science fair projects having taken place in 1956. However, the authors observe that in recent years, there has been a dip in the quality of the science fair projects conducted by primary school students. They explain this in terms of teachers lacking expertise in inquiry-based learning due to a lack of experience in doing inquiry themselves, the lack of time, and a shortage of resources. Another factor would appear to be that summative assessments focus primarily on testing conceptual understanding in science, with little or no test items that address process skills of inquiry learning. Further to this, due to teachers' lack of a background in inquiry learning, they find it a challenge to develop assessments that evaluate students' hands-on or process skills as well as inquiry ability. This is in sharp contrast to China where assessment tasks of inquiry learning are specified for teachers.

In Korea, the 2015 national curriculum aims to develop inquiry skills so that learners may solve problems of natural phenomena and daily life scientifically. Science inquiry and science competences for critical and creative problem-solving are noticeably emphasized in the 2015 national science curriculum. This emphasis requires less learning of science content and instead has more inquiry practices in school. Throughout the history of curriculum development in Korea, it is evident that science inquiry has taken on increasing importance. As is the case with other East Asian countries, there would appear to be a disconnect between the specifications of the formal curriculum and the enacted curriculum. The authors cite research studies that show that teachers have difficulties in the enactment on inquiry-based instruction. Teacher belief would appear to be a factor on the implementation of inquiry pedagogy with calls for research on teachers' perceptions of the changes of policies and reflections on changes in their actions (Hall and Hord 2011; Kim and Tan 2011). Although Korean learners are top achievers in TIMSS, the TIMSS results also reveal that students' interests and motivation in science are low. The authors point out this issue is addressed in the new science curriculum 2015, where "inquiry-based and performance-based learning and teaching are encouraged to enhance students' interests and curiosity." Here, inquiry activities are suggested to help teachers and students achieve achievement standards more effectively. Yet, the authors maintain that important learning objectives such as interest and curiosity toward science, scientific attitudes, and the joy of learning science should be more explicitly addressed in the curriculum. The importance of inquiry skills as a core teacher competency is affirmed through the National Teacher Employment Test that teachers much pass in order to teach in public schools. Here, teachers are tested on inquiry skills.

The preceding discussing on the primary science curriculum of Singapore, Taiwan, China, Japan, Hong Kong, and Korea as elucidated by the respective chapter authors has affirmed the pivotal role of the teacher in curriculum reform. Bybee (1997) maintains that if teachers' practices do not represent the curriculum innovations, the entire process of curriculum change falters and eventually fails. Indeed, the teacher can be regarded as an agent of curriculum change. In all six countries, we learn that despite good educational intent for a shift toward inquiry-based primary science education, this pedagogy has not gained classroom traction to the extent to which it was envisaged. It is widely acknowledged that inquiry-oriented teaching places high demands upon teachers' conceptual literacy because it plays a pivotal role in facilitating student learning during inquiry-oriented teaching. According to Mines (1995), "the art of skilful questioning appears to be crucial to achieve the balance between giving students suitable guidance and leaving sufficient scope for then to think independently" (p. 14). This can only be achieved if teachers' conceptual literacy is sound.

Colburn (2000) places much responsibility for the success of inquiry-based learning on the teacher, by emphasizing that teachers are the central decision-makers in regard to the way in which an inquiry approach is construed and adapted in the classroom. According to the NRC (2005), in a class where inquiry is taking place, the teacher requires a deep understanding of scientific processes in order to guide learners in formulating an investigation question and planning the investigation, as well as a deep understanding of science concepts in order to guide them toward subject matter understanding and other learning goals. If this is lacking, it does contribute to the unsatisfactory experience of school science which learners commonly have (Kim & Tan 2011). Crawford (2000) also suggests that teachers' content knowledge, pedagogical content knowledge, and theoretical knowledge must act in concert before teachers can implement inquiry.

This then raised the question of the adequacy of pre-service and in-service teacher education in the East Asian countries that form the focus of this book.

For Singapore, the authors highlight the importance of teacher training toward high national achievement in TIMSS. The authors relay that "capacity building is both abundant and frequent for primary science teacher in Singapore." They provide a detailed description of courses that are offered for both pre-service teachers who seek certification and in-service teachers for continuing professional development. The professional development efforts would appear to be in sync with curriculum reforms. For example, at the pre-service level, in the course *Curriculum and Practices for Primary Science*, beginning teachers receive instruction in the nature of science and scientific inquiry, which are regarded as key principles of science teaching and learning in the reformed curriculum. At the in-service level, the NIE Advanced Diploma in Primary Science Education encompasses modules on teacher pedagogical content knowledge and inquiry skills. NIE also offers short courses that practicing teachers can take to be incorporated into their personal PD plans for the school year. These courses have much relevancy for teachers in implementing a reformed science curriculum and include topics such as STEM education, scientific argumentation, discourse studies in science, case-based science investigations, problem-based learning, modeling instruction, as well as topics in the natural sciences. In addition to professional development at NIE, the Academy of Singapore Teachers (AST), which is run by the Ministry of Education, further provides professional development for all teachers across subjects. Here again, there are initiatives that support teachers in driving a progressive science pedagogy. A subject chapter in primary science led by master teachers is formulated to "equip Primary Science teachers with impactful and innovative pedagogies that are underpinned by sound learning theories, experimentation and research." It is also interesting to note that experienced teachers conduct short workshops for parents on how they could support their children in the inquiry-based learning at home.

In Japan, teachers constitute a learning community, and accordingly they are regarded as being at the forefront of innovative science teaching. They are frequent contributors of articles and papers that are published in journals such as Science Education Monthly. This supportive and collaborative learning is also facilitated by teachers who publish books that include new teaching and learning materials. Often suggested lesson plans are included.

In China, with the goal to achieve equity and core competencies, constant reforms in curriculum and teaching have been the norm. Teacher development programs have been prioritized for 2020 so that teachers' professional and academic level can be improved. Four modes have emerged in training pre-service primary science teachers with undergraduates coming from elementary education universities, natural science departments of education universities, science departments of universities, and colleges for teacher training. Each mode reflects "different routes of curriculum reform in elementary education." Professional support for in-service teachers in implementing curriculum reforms is divided into two stages, namely, centralized training and decentralized training. Centralized training is comprised of lectures and seminars, participatory training, case-oriented training, and workshop-oriented training. Of particular note here is the participatory training where teachers interact with each other in an enabling environment that supports them in exploring their beliefs on innovative practices. They also acquire firsthand experience of the innovation by doing. For example, teachers can experience the process of scientific inquiry by doing inquiry. The authors maintain that through "these forms of active participation in an atmosphere of scientific inquiry, teachers' teach ing ability and level will gradually improve." The participatory nature of this professional development is also decentralized when it is extended to the school level where teachers put into practice what they have acquired at the centralized level. Teachers are encouraged to communicate with each other in sharing ideas and exchanging views about teaching design, teaching philosophy, and teaching strategies.

In Hong Kong, major curriculum reforms have been met with obstacles. Teachers have struggled in making the transition in pedagogy from the content-focused approach toward a learner-focused approach that targets the development of inquiry and investigative skills. The factors stifling curriculum reform are teacher related and include teachers lacking subject knowledge, relevant experience, and pedagogical knowledge.

In preparing students teachers to be adept at teaching General Studies, a subject that include science, there are initial teacher education programs, including the Bachelor of Education (BEd) for primary pupils and the Postgraduate Diploma of Education (PGDE) for students who already possess a degree. In Hong Kong, there is also strong focus on the ongoing profession professional development of in-service teachers so that they may ready to address challenges that result from curriculum reform initiates. Certainly, the teacher is cast in the role of a lifelong learner. It is a requirement that every teacher has to complete 150 h of professional development activities in a 3-year cycle. The authors report that in supporting teachers to acquire competency in inquiry, pedagogy workshops are offered "to equip teachers with some basic scientific understanding and provide them with hands-on opportunities for designing some practical science activities."

Due to the priority that is given to teacher professional development in primary science, it is not surprising that this occupies a significant place in the research agenda of science education research in Hong Kong. The research has centered mainly on investigating changes in teacher pedagogy and orientation over a period of time. For example, So and Watkins (2005) conducted a longitudinal study to explore how 25 teachers' thinking changed from pre-service teacher education through to their first year as classroom teachers. The researchers found that the majority of the teachers became more constructivist in terms of their conceptions and teaching practice, but they also tended to become more simplistic in planning and less coherent in thinking as they progressed from pre-service teacher education to beginning teachers in schools. Research has also been conducted on the efficacy of variation theory as a guiding principle of pedagogical design when adopting a lesson study approach.

In Taiwan, primary science teachers are expected to design suitable inquiry-based learning activities for their students by themselves. However, these efforts are constrained due to teachers lacking competency of inquiry learning. The authors flag that a major impediment in the professional development of pre-service primary science teachers in Taiwan is that it is not a requirement for students to take science courses in their teacher training. As a result, graduates do not have sufficient pedagogical content knowledge for effective science teaching. The authors also highlight that during the internship program, these pre-service teachers do not get the practical experience of teaching science because the emphasis during school practicum is predominantly on classroom management skills and the two major courses, mathematics and Chinese.

As is the case with Hong Kong, the continuous professional development of Taiwanese primary school teachers is regulated. The Ministry of Education stipulates that primary school teachers need to attend further studies for 18 h every semester year or accumulate 90 h (5 credits) in a 5-year period. However, it is noted that science education workshops only account for just a small part of this ongoing development. The primary means of professional development for in-service science teachers is mentor groups that are established by local governments. This is a viable form of professional development and has increased in popularity worldwide as a way by which a teacher experiencing some weakness in his/her practice can be supported by a skilled and experienced colleague (Bradbury 2010). In these mentor groups, excellent and experienced primary science teachers act as mentors to novice teachers in supporting them improve their science teaching competency. The success of these mentor groups is recognized by the Ministry of Education with mentor groups now operating at the national level. Mentor groups now identify schools where teachers are in need of professional development and demonstrate science learning activities such as inquiry activities. Apart from providing in situ mentoring, mentors are also invited as lecturers during Wednesday afternoon workshops. Science teachers also are involved in online learning communities such as *Facebook* that allows them the convenience to discuss and share ideas at a time that is convenient. The authors also report that the Ministry of Education in Taiwan actively encourages science teachers to conduct science education research in order to enhance their teaching practice. This research is mainly in the form of action research and is effective in enabling teachers address their practical teaching problems associated with curriculum reform. This form of research has been deemed to be effective in improving professional development, and teachers also collaborate in sharing their results. It would therefore appear that in Hong Kong, action research provides teachers with new knowledge and understanding about how to improve their teaching practices and to resolve significant curriculum problems in their classrooms.

In Korea, pre-service teacher education students are admitted to a 4-year program at a university. Quality of teacher output is underlined with teaching certificate holders having to pass the National Teacher Employment Test in order to teach in public schools. The National Teacher Employment Test is administered annually by the Korea Institute for Curriculum and Evaluation, a government-funded educational research institution, and each Metropolitan and Provincial Office of Education. Upon passing this test, teachers gain tenure to teach in public elementary schools; thus, this test is extremely competitive and challenging.

The teacher education curriculum in most of the universities includes two to three times of school practicum during 9–10 weeks in total. Usually every student teacher teaches 15–20 class hours during 4 weeks of the practicum; thus, they may teach one to three science lessons for the entire practicum. This is not enough to develop pedagogical strategies and confidence in science teaching, especially that pre-service teachers showed great difficulties in conducting science practical work in classroom teaching and found the gap between university instructions and actual teaching with science experiments (Yoon 2004). This challenge is also closely relevant to the issue of elementary teachers as generalist versus specialist. Like other countries, elementary school teachers in Korea usually teach most of the subjects; thus, they go through generalist tracks. This means there is not enough time to develop science knowledge and to practice science teaching during practicum. The debate on specialist and generalist in elementary education is still ongoing, and this issue is flagged as being critical to the quality of science teaching and the effective delivery of the primary science curriculum.

This commentary shows that although the challenges that each country faces are different, there are several commonalities and shared interests. In this regard, I have decided to zoom in on two key issues, namely, strides that these countries are making from a content-based curriculum toward a curriculum that has a strong inquiry-based focus. This shift has resulted in difficulties teachers experience that emanate mainly from their lack of preparedness. The education ministries of each country address this concern by putting in place strategies to support practicing teachers. We also see in teacher education at the pre-service level university curricula that attempt to address preparedness of teachers to implement curriculum reform.

References

Abrahams, I., & Millar, R. (2008). Does practical work really work? A study of the effectiveness of practical work as a teaching and learning method in school science. *International Journal of Science Education, 30*(14), 1945–1969.

American Association for the Advancement of Science. (1993). *Benchmarks for science literacy*. New York: Oxford University Press.

Bartos, S. A., & Lederman, N. G. (2014). Teachers' knowledge structures for nature of science and scientific inquiry: Conceptions and classroom practice. *Journal of Research in Science Teaching, 51*(9), 1150–1184.

Bradbury, L. U. (2010). Educative mentoring: Promoting reform-based science teaching through mentoring relationships. *Science Education, 94*(6), 1049–1071.

Bybee, R. (1997). *Achieving scientific literacy: From purposes to practices*. Portsmouth: Heinemann Educational Books.

Castells, M. (2000). *The information age: Economy, society and culture* (Vol. 3). Oxford: Blackwell.

Chinn, C. A., & Malhotra, B. A. (2002). Epistemologically authentic inquiry in schools: A theoretical framework for evaluating inquiry tasks. *Science Education, 86*, 175–218.

Chiu, M. H. (2007). Standards for science education in Taiwan. *Making it comparable: Standards in science education* (pp. 303–346).

Colburn, A. (2000). An inquiry primer. *Science Scope, 23*(6), 42–44.

CPDD. (2007). *Science syllabus primary 2008*. Singapore: CPDD/Ministry of Education.

Crawford, B. (2000). Embracing the essence of inquiry: New roles for science teachers. *Journal of Research in Science Teaching, 37*(9), 916–937.

Department for Education and Employment. (1999). *The national curriculum for England: Key stages 1–4*. London: Qualifications and Curriculum Authority.

Hall, G. E., & Hord, S. M. (2011). *Implementing change: Patterns, principles, and potholes* (3rd ed.). Boston: Pearson.

Hodson, D. (2003). Time for action: Science education for an alternative future. *International Journal of Science Education, 25*(6), 645–670.

Kim, M., & Tan, A.-L. (2011). Rethinking difficulties of teaching inquiry-based practical work: Stories from elementary pre-service teachers. *International Journal of Science Education, 33*(4), 465–486.

Lead States, N. G. S. S. (2013). *Next generation science standards: For states, by states*. Washington, DC: The National Academies Press.

Lee, Y. C., & Ng, P. H. (2004). Hong Kong primary pupils' cognitive understanding and reasoning in conducting science investigation: A pilot study on the topic of "keeping warm". *Asia-Pacific Forum on Science Learning and Teaching, 5*(3), Article 8.

Liu, X.-f., Liang, L.-L., & Liu, E.-S. (2012). Science education research in China: Challenges and promises. *International Journal of Science Education, 34*, 1961–1970.

Mines, G. (1995). Sc 1 investigations: Management and assessment at key stage 4. *Education in Science, 165*, 13–15.

Ministry of Education, Culture, Sports, Science & Technology. (2006). *Images of revising learning content of science* (draft), Central Council for Education, Primary and Secondary Education Sectional Committee, Curriculum Subcommittee (fourth stage, 11th) conference minutes/documents distributed 4-2. [In Japanese].

National Research Council. (1996). *National science education standards*. Washington, DC: National Academy Press.

National Research Council. (2005). *America's lab report: Investigations in high school science*. Washington, DC: The National Academy Press.

National Research Council. (2012). *A framework for K-12 science education: Practices, crosscutting concepts, and core ideas*. Washington, DC: National Academies Press.

Osborne, J. (2000). Science for citizenship. In M. Monk & J. Osborne (Eds.), *Good practice in science teaching* (pp. 225–240). Milton Keynes: Open University Press.

Poisson, M. (2000). Science education for contempory society: Problems, issues and dilemmas. Final report of the international workshop on the reform in the teaching of science and technology at primary and secondary level in Asia: Comparative references to Europe. http://www.ibe.unesco.org.

Rogan, J. M., & Aldous, C. (2005). Relationships between the constructs of a theory of curriculum implementation. *Journal of Research in Science Teaching, 42*(3), 313–336.

Roth, K., Chen, C., Lemmens, M., Garnier, H. E., Wickler, N. I. Z., Atkins, L. J., Calabrese-Barton, A., Roseman, J. E., Shouse, A. W., & Zembal-Saul, C. (2009). *Coherence and science content storylines in science teaching: Evidence of neglect? Evidence of effect?* Symposium on *Science content storylines* presented at NARST meeting, Garden Grove, Orange County CA, April 17–21, 2009.

So, W. M. W. (2013). Connecting mathematics in primary science inquiry projects. *International Journal of Science and Mathematics Education, 11*(2), 385–406.

So, W. M. W., & Ching, N. Y. F. (2011). Pupil science learning in resource-based e-learning environments. *Journal of Computers in Mathematics and Science Teaching, 30*(2), 203–223.

So, W. M. W., & Watkins, D. (2005). From beginning teacher education to professional teaching: A study of the thinking of Hong Kong primary science teachers. *Teaching and Teacher Education, 21*(5), 525–542.

So, W. M. W. (2006). Inquiry into primary pupils' science projects: Implication for design of inquiry learning. *Asia-Pacific Forum on Science Learning and Teaching, 7*(1), Article 3.

So, W. M. W., Cheng, N. Y. I., Chow, C. F. S., & Zhan, Y. (2014). Learning about the types of plastic wastes: Effectiveness of inquiry learning strategies. *Education, 3-13*, 311–324.

Song, Y. (2014). "Bring your own device (BYOD)" for seamless science inquiry in a primary school. *Computers & Education, 74*, 50–60.

Song, Y., & Kong, S. C. (2014). Going beyond textbooks: A study on seamless science inquiry in an upper primary class. *Educational Media International*, Online publication, 1–10.

United States Department of Commerce, United States Department of Education, United States Department of Labor, National Institute of Literacy & the Small Business Administration. (1999). *21st century skills for 21st century jobs*. Washington.

Yeung, S. Y. S. (2009). Is student–centered pedagogy impossible in Hong Kong? The case of inquiry in classrooms. *Asia Pacific Education Review, 10*, 377–386.

Yoon, H. G. (2004). Science education using science drama. Proceedings of the 46th conference of The Korean association for research in science Education, 26–43.

Chapter 11
Curriculum, Pedagogy, Teacher Training and Recent Reforms in Primary Science Education

Lorraine Pe Symaco and Esther G. S. Daniel

11.1 Development of Primary Science Education

The role of education for the development of countries has reorientated government policies worldwide to focus and increase investments in their education sectors. With the rising trend of internationalization and the relevant knowledge-based societies, there has been an increasing demand to ensure that school and university graduates are able to adapt to the needs of the modern times. The neoliberal orientation of the broader education sector has also witnessed the rise of international testing assessments and rankings such as the Trends in International Mathematics and Science Study (TIMSS) and the Program for International Student Assessment (PISA), among others. In line with this, the more recent focus in science, technology, engineering, and mathematics (STEM) has further highlighted the role of the sciences in wider socioeconomic advancement of counties. Additionally, general issues on curriculum, pedagogy, and teacher training in science education have pushed the discussion forward in terms of finding the most appropriate approach to guarantee that students receive the best training in the subject.

The following sections will broadly review issues relating to curriculum and pedagogy, teacher training, and recent reforms in science education.

L. P. Symaco (✉)
Zhejiang University, Hangzhou, China
e-mail: lorraine@symaco.org

E. G. S. Daniel
University of Malaya, Kuala Lumpur, Malaysia

Y.-J. Lee, J. Tan (eds.), *Primary Science Education in East Asia*, Contemporary Trends and Issues in Science Education 47,
https://doi.org/10.1007/978-3-319-97167-4_11

11.1.1 Curriculum and Pedagogy

The increasing influence of globalization has instigated the need to reform and present a more relevant science curricula (Carter 2005). From a content-based to a more contextual approach to the sciences (see Childs 2015), the varied curriculum development of science education over the years points to the anticipated and desired direction of the subject which will enable students to utilize the field as appropriate to the needs of the changing society. The pivotal meaning to a curriculum emphases of course cannot be overemphasized when "one of the reasons for the different science curricula over time or the differences between countries or between levels of education, is a difference in what are considered to be the important curriculum emphases" (Childs 2015, p. 387).

Different approaches to curriculum development which range from utilizing practical and thinking skills, hands-on investigations to underlining technological and scientific literacy illustrate the diverse contextual actualities of science education in schools. The highlight on education for sustainable development has also influenced the broader science education curricula through inclusion of more environmentally conscious courses for students especially at the school level. The connection to science education is evident as Van Eijck and Roth (2012) describe this further where,

> [a] call for a science education that brings about and maintains a high level of scientifico-scientific literacy in the general population. In the chain from the current (environmental) problems towards solutions that contribute to a sustainable future of life on Earth, including humanity as a whole, science education is increasingly recognized as a critical link. (p. 165)

This vital role of science education can perhaps be best exemplified on how children often transmit what they learn in schools to their immediate familial connections, which can prove to be crucial especially for families or households that have a generally low completion level of formal education. The heroic case of Tilly Smith then at age 10 and her role in the 2004 Indian Ocean earthquake demonstrate the powerful role of education and its significant influence in children. Recalling the signs of tsunami formation from her geography class, she was able to warn people before its actual occurrence, thus saving nearly a hundred lives in a resort in Thailand on that disastrous day. Similarly in this chapter, country case studies on how curriculum developments across China, Hong Kong, Japan, Korea, Singapore and Taiwan show how respective nations develop, modify and improve their science primary education curricula as relevant to their contextual realities and to the wider global concerns.

Related to the development of curriculum are actual pedagogical approaches to primary science education in the covered country case studies. The focus on a child-centered approach in pedagogical style (as opposed to teacher-centered) has over the years demonstrated the tenacious hold of this method over others in view of its benefits. This has been given much credence to effective teaching and learning processes where in a child-centered education, "the curriculum begins with the needs and interests of the child and responds to the unique characteristics of childhood.

Teachers use their knowledge of how children develop to structure learning experiences that facilitate children's learning through play and discovery" (Ryan as cited in Langford 2010, p.114). But despite some criticisms to the child-centered approach, among others, as semblance of the developmentally appropriate practice (DAP) (see Brooker 2005) and the decontextualized construct of the free and individual child (see Langford 2010), advocates of this pedagogical practice remain.

This line along with inquiry-based teaching and learning, experimentation, and outside classroom activities is but some of the active pedagogical approaches featured in the relevant country case studies reviewed in this chapter. Orientating children on the realisms of science through outside classroom activities, cultural exchanges, and visits offer unique perspectives to students of the science "at work," which are often lacking in the traditional classroom setting. Given this, a more liberal, engaging, and hands-on experience is seen to exemplify the recent approach to teaching of primary science education in schools.

11.1.2 Teacher Training and Recent Reforms

While the pedagogical approach used in primary science education is a major issue, another equally important factor to consider is the actual training of science education teachers. The debate between content/subject matter versus pedagogical knowledge stems from the often perceived lack of content expertise of teachers in the sciences. As will be expounded later in this chapter, this has pushed for a more accessible and open teaching practice wherein science majors (as opposed to general education majors) are allowed to teach in schools. This evidently also shifts the concern where the latter are considered lacking in pedagogical content which can then result to the students learning far less than intentioned. Literature has shown this prevalent concern (e.g., Loughran et al. 2015; Zhang and Birdshall 2016) and advocates for a balanced approach is expected, where Brynes mentioned this crucial feature:

> insofar as a teacher's knowledge provides the basis for his or her effectiveness, the most relevant knowledge will be that which concerns the particular topic being taught and the relevant pedagogical strategies for teaching it to the particular types of pupils to whom it will be taught. (as cited in Darling-Hammond 2000, p.167)

Additionally, the concern of teacher training internships to make available a more contextualized account of the classroom setting to teacher trainees is another issue exhibited in some of the countries case studies. Given such concerns, the significance of teacher education for effective teaching and learning in the classrooms is undeniable where teachers with training in teaching are found to be more "confident and successful with students than those who have had little or none" (Darling-Hammond 2000, p.161).

Some of the recent reforms in science education in the country case studies vary in form, from increasing quality in science education through improved teacher training and resource mobilization to the increased assessment and rankings culture as manifest in national and international assessment schemes as mentioned earlier (e.g., TIMSS and PISA). The rise of the knowledge-based society has also elicited a response in the sector which incorporates features of information and communications technology (ICT) in delivering science education in schools. Equally relevant of course is the focus of governments in STEM which hopes to equip a country's human capital with the needs of the modern world. Subjects such as those featured in STEM are projected to assist in providing the needed technical expertise required in the k-economy. Notwithstanding such advances in the global society, a more conscious effort to promote sustainable development is also realized in the wider education curricula worldwide where existing "school systems and teacher trainings" are tailored to meet the challenges of a changing world (Battro et al. 2017).

The following sections will provide a synoptic review on the curricula, pedagogy, teacher training, and recent reforms in primary science education in China, Hong Kong, Korea, Japan, Singapore, and Taiwan.

11.2 Science Curricula

Science education in many Asian countries took shape in a formal way after World War II in 1945. The curriculum was further influenced when the USA in 1957 surprised the world with Sputnik 1. Many countries went back to the basics of science and scientific skills, where scientific content became prominent. The primary science curricula in Korea, Japan, China, Hong Kong, Singapore, and Taiwan were no exception.

11.2.1 History

The science curriculum in Korea began to crystallize after independence from Japan in 1945, very much influenced by the afterwar progressivism perspective. When America successfully sent the first man into space in 1957 with a successful lift-off of Sputnik 1, science curricula in Korea became more structured in terms of the knowledge to be learnt and instructional strategies, leading to an overload of science content. Nevertheless, since the 1980s the science curriculum has taken on a more holistic development of students' perspective. As in Korea, primary science education in Japan began in 1947 after World War II and has been revised eight times since then. In the early years, systematic scientific content was emphasized, which slowly changed due to the by scientific advances in the USA after Sputnik 1 in 1957. Step by step, discovery and inquiry learning became more prominent, and since the 1970s, environmental issues came to the forefront to ensure familiarity and closeness to nature.

Compared to other Asian nations, China faces enormous challenges because of its vast land area and much diversity in culture and ideology. In spite of this, the science curriculum is developed centrally (2012), and the first science curriculum standards were put in place in 2001, the first since the communist take-over in 1949. As the economic standing of China continues to rise, there continues to be a move from the elite to a "science for all" where scientific literacy is being emphasized (Wong et al. 2014). In comparison, primary science education was introduced in Hong Kong in the 1970s, which was previously known as "Nature Study" in the 1960s, while a primary science curriculum to help learners discover and experience scientific truths through the natural environment was put in place from the 1980s. Since then, much research on its implementation has given rise to continued improvement and enhancement of the curriculum in Hong Kong.

The passage of Singapore science education in history is unclear, but it has its roots in the Malaysian system till 1963. However, what is clear is that since the 1990s where the focus was on knowledge and facts of science, Singapore has come a long way in just over two decades to become a formidable force in science education. Emphasis is on skills and preparation of their students for practice in professions in scientific fields. Conversely, Taiwan's science education stems from the setting up of a scientific education committee and the establishment of the National Taiwan Science Museum in 1954, which were both responsible for the counseling and development of scientific education matters. In 1962, the National Taiwan Science Museum was renamed as National Taiwan Science Education Museum and became the focal point in promoting science education and in improving the overall scientific literacy of the nation as a whole (Zhang Lai Miao Li 2000).

Clearly, each nation has its own history of science education and how it evolved, especially after World War II which was a crucial turning point. However, as the world began to face the environmental crisis since the early 1960s, science curricula have slowly changed to address the problems brought on by anthropogenic activities.

11.2.2 Present Directions

In present times, the science curriculum gives priority to three components, namely, the facts and concepts of science (content), the nature and processes of science (conduct or process), and the applications of science in society (context) (Childs 2015). The current curricula of all the six nations lean toward creating scientifically literate citizens through these three components. In Korea, this means expanding from just content knowledge to building practical skills and thinking skills to make decisions in solving everyday problems in an ever increasingly complex world. The content is integrated from learning about nature and natural things to the physics of sound, acids, and bases. Inquiry processes are given importance for improving curiosity and developing lifelong skills to deal with present-day ecological and biotechnological challenges.

Embracing these very same ideas, the primary science curriculum in Japan also has an integrated curriculum of natural and physical science components, with an emphasis on the study of the living environment. It is obvious that the goal of elementary science curriculum in Japan emphasizes the role of the subject in terms of understanding living things and human influence upon nature, leading to ideal attitudes that respect the environment. Hands-on experimentation is given credence for the acquisition of problem-solving abilities.

The science curriculum of China is not far off from the Korean and the Japanese curricula. Besides the conceptual content, the affective aspects of values and positive attitudes are given priority where the learners are allowed the freedom to be researchers in the classroom and to connect what they learn with everyday situations. In other words, besides the content of science and the scientific skills that the students need to acquire, the affective aspects, attitudes, and values are also given the same level of importance. Similarly, Hong Kong's science curriculum content also integrates both natural/physical and earth science issues. It also involves hands-on investigations that lead learners to a better understanding, better skills, and appreciation of the world around them.

Singapore distinctly embraces the making of competent future professionals who will possess the skills infused from a young age, to face the challenges related to social situations and the environment. The natural and physical science content of Singapore is similar to other country case studies but is further integrated into five themes which are diversity, interactions, systems, cycles, and energy, with emphasis on the inculcation of ethics and positive attitudes. Finally, Taiwan is the epitome of all the rest but explicitly states the role of technology and scientific literacy in the solving of everyday problems. In particular, the end objectives for teaching science is to propel its citizens to explore and understand scientific knowledge in order to apply this understanding in addressing daily challenges faced by the nation, especially in the drastically changing natural environment.

Given science education's seeming contribution to global citizenship through the nurturing of children's interest in the world around them and developing positive attitudes toward science (New Zealand Council for Educational Research (NZCER) 2010), the curricular content of the subject is essential. Overall, the country case studies are moving in the same direction, in that the science curricula content have been revised to emphasize the scientific skills to be acquired in order to face this challenging era in finding solutions to global issues.

11.3 Pedagogies

Pedagogy in the primary science classroom is important in order to critically engage students' thinking; otherwise, the tendency for student to disconnect and apply rote memorization is likely. The different pedagogies discussed in this book across the six nations indicate that efforts are taken to ascertain that science learning goes beyond the mere familiarization of content, to actual application of skills and knowledge to everyday challenges. In terms of pedagogies/teaching, it can be said that all the six nations have major shifts toward inquiry teaching and learning. Every country case study reports a shift from teacher-centered instruction to child-centered learning, which involves encouraging problem solving, creative thinking, learning to learn, and activity-based classroom and out of class programs (National Inst. for Educational Research, Tokyo 1999).

Experimentation and inquiry appear to be the current words that best describe prevalent science teaching approaches across the six nations. In China, the inquiry approach is taken seriously, such that students actively participate in the cognitive processes of problem identification and problem-solving rooted in evidence and data interpretation. In addition, science activities outside of the classroom is apparently frequent, though at its beginning stage, where students are exposed to and participate in various events such as science based exhibitions, workshops, competitions, and camps. To a certain extent, the inquiry approach in Hong Kong is also similar to that of China, in that students are involved in outside classroom activities such as visits to the science museum, workshops, and camps. Nevertheless, elementary science education in Hong Kong has gone a step further as it has become the focus of much classroom research in generating a theoretical model to conceptualize the processes of scientific inquiry. Another aspect in Hong Kong's science education is the inherent use of technology in the teaching-learning process.

In the case of Singapore, it has also been found out that class learning such as visits to the science center and the cryogenics camp has increased students' enjoyment and interests in science, similar to China and Hong Kong. In addition, these visits are planned with giving students directions for their future professional careers. Inquiry in the science classroom has evolved from one of just doing science via experiments to a deeper form of evidence-based reasoning and application to everyday problems. Additionally, Singapore science teachers use a range of teaching techniques from games, drama to demonstrations, and project work skewed toward the 5E stepwise approach to inquiry learning.

To develop positive attitudes toward science, Taiwan similarly advocates outdoor science learning within the school environment as well as in natural settings around the island. Cultural exchange activities related to science education are encouraged. Science fairs are also popular to instill interest in learning the subject. To develop scientific thinking and students' science process skills, Taiwan highlights the inquiry approach as well. A unique component in Taiwan's science education scenario is reading, where students in school (and the public at large) are directed to better scientific literacy by reading various printed media related to science.

Science education in Korea also places importance on scientific thinking and inquiry skills. Korean science educators have been found to be above the TIMSS international average in being able to engage students in their lessons, although they have been found to be less confident in the classroom compared to other nations. The heavy reliance of teachers on the standardized science textbooks is also noted in Korea, which reflects the lack of creativity of teachers in terms of lesson planning. This is perhaps telling of the need to develop further the pedagogical skills of Korean science teachers.

It is well accepted that effective science learning is achieved through direct experience and experimentation, where students' motivation, powers of reasoning, problem-solving abilities, critical analysis, and creative thought are increased (Huang 2012). With the aforementioned pedagogical methods observed in the country case studies, it is promising to see that effective science learning is a prime objective to all.

11.4 Teacher Learning and Training

Teacher learning and training in China, Hong Kong, Japan, Korea, Singapore and Taiwan generally take an identical vision, that is, to ensure that students in schools get the most qualified and competent science teachers. This is to make certain that students will be ready to face the challenges of the knowledge society when they leave school. While teacher training and qualifications usually take place in specialized teaching institutions (e.g., teacher normal schools), government policy reorientations in certain countries have resulted in teacher education programs being offered alongside other university courses. For instance, the distinct case of Korea which offers a 4-year elementary education program highlights the advantages of offering specialized skills and knowledge at the entry level while at the same time making it complicated to change career tracks in the middle of the program. Similarly, highly specialized teaching institutions existed in Taiwan prior to the Teacher Education Act of 1994. Currently, universities in Taiwan also offer teacher education programs, though criticisms of a qualitative science teacher training exist. On the one hand, countries like China offer a number of routes for teacher training (e.g., the four "modes"), which is in direct contrast to Singapore's only teaching training institution, the National Institute of Education (NIE).

Professional qualifications and teacher training also vary, from an assessment licensure-type-based examination in Korea and Taiwan to qualifications by degree in Hong Kong. Internships and professional placements to complement teacher training courses are also a common practice. Such training placement also exposes pre-service education students to the contextual realities of the classroom and allows them to bridge the "theory to practice" gap, highlighting further the benefit of such placements (Goodlad 1990, McNally et al. 1997). In Singapore and Korea, a credit of about 10 weeks is allotted for classroom/school placements, and similarly a full semester is allotted to such exercises in Taiwan. Though for the latter, such placements are criticized for their focus on classroom management skills, with little to nil actual science classroom teaching opportunity. In Korea, students are also exposed to special education schools during such training.

Cannon (1997) also highlights specifically the role of extended science practicum experiences in improving pre-service primary science teachers' (science) teaching efficacy beliefs and where the acknowledged impact of practical teaching brings about such wide benefits on professional and personal development of pre-service teachers:

> The diversity and complexity of settings, interlocutors and activities that characterise this process are responsible for what we could consider a real 'personal revolution', through which the neophytes have the opportunity to discover themselves in the role of teachers, to interact with a new set of situations, to practise and play new adult and professional roles and tasks, or to acquire, develop and/or consolidate new skills and knowledge. (Caires and Almeida 2005, p. 117)

Another common issue surrounding teacher education training is the perceived and often acknowledged lack of specialized content knowledge of teachers in their areas of teaching. While this can range from actual limited training during their pre-service studies due to a generalist orientation, or when science teachers are not actual science majors, the issue of "generalist" versus "specialist" broadly defined comes into context. While some countries are opening their schools to actual science degree majors (in contrast to education degree majors), the issue of further training the former in pedagogical techniques in the classroom is another thing to consider. Such is the case in Korea, where policy realignments are sought to allow graduates from institutions other than the teacher training colleges to teach science in schools. Hong Kong similarly faces the same setback wherein the majority of teachers trained under the general studies route have non-science backgrounds, which makes it more difficult for such teachers to respond to student inquiries and clarify science concepts.

Singapore addresses this issue by deploying teachers who are specialists in the field, which will then complement the subject heads and heads of departments in schools. Despite such contention, the importance of both subject matter and pedagogical knowledge in effective teaching and learning is emphasized in literature. Whereas solely relying on subject matter knowledge is insufficient for effective teaching, a "critical amount of subject matter knowledge seems to be necessary to develop the pedagogical content knowledge required (...)" (Magnusson et al. 1999, p.123). Given this, a number of initiatives that will further develop science teachers in the country case studies are exhibited. While professional development may vary across countries, the recognized need to further train primary science teachers is evident.

Various initiatives and requirements cover further professional development of science teachers in schools. Some case studies discussed here focus on the need to upgrade teaching in rural settings, such as China's National Training Plan (NTP), while others require compulsory professional development beyond in-service teacher training. For instance, Hong Kong requires their science teachers to complete a 150-h professional development activities in a three - year cycle, while Taiwan requires a further study of 18-h per semester, or alternatively earn 90-h in a five-year period. In Hong Kong, support is given by the Education Bureau for science teachers to meet such requirements, though the actual practice of learned pedagogy in such occasions is rarely observed given the ad hoc nature of such activities. Hong Kong's Chief Executive's Award for Teaching Excellence (CEATE) in Science Education also serves as a venue to further professional development where awardees receive a continuing professional development scholarship. This award also allows teachers to share teaching good practices among themselves. Integrated programs for professional development are also practiced, such as the case in Taiwan wherein it aims to "integrate different hierarchical educational organizations." The Ministry of Education of Taiwan also sponsors an annual science education fair which, among others, serves as a venue to exchange educational research and good practices.

Professional development is also much encouraged in Singapore where on average, 98% of teachers have participated in such activities, which exceeded OECD norms (in 2013). In Singapore, professional development courses are offered by the Academy of Singapore Teachers (AST), while further studies and a variety of ad hoc activities for further development are also available at the NIE. The popular Enhanced Professional Development Continuum Model offered by NIE since 2012 allows graduate teachers to pursue a 2- to 4-year master's degree in various disciplines, with support from the Ministry of Education. Teacher contributions to innovations in science teaching and learning are also considerably supported in Japan. Supplementary professional development is advocated by the Society of Japan Science Teaching (SJST) where science teachers are encouraged to share innovative practices through the *Science Education Monthly* publication.

11.5 Recent Reforms in Science Education

One significant reform mentioned in the case study chapters includes the revision of national curricula, which naturally affects the science curricula in schools. While a number of countries revise their national curriculum every few years, such as Korea's revised national curriculum to be implemented in 2017 and Singapore's revision every 6 years, the contextual realities inherently unique in each case study dictate the manner to which science teaching and learning are assumed. For instance, despite the economic superpower that China is, their National Training Plan of 2010 as mentioned earlier focuses on the need to ensure that their rural primary science teachers are competently trained in the subject area. This reflects the uneven development in China where despite the existence of top performing schools, the government needs to make certain that those within the periphery are also able to acquire the needed training and development for success.

The increasing emphasis on science achievement has also paved way to countries directing efforts to alternative and/or complementary ways of learning. In China, recent reforms include the experimental "learning by doing" (LBD) which was launched 15 years ago which aims to offset the "reality of disadvantages present in education in China." This LBD program is promoted by both the Ministry of Education and the China Association for Science and Technology. Additionally, local programs that highlight innovative teaching practices in the sciences are observed, such as the "period of integrated studies" in Japan which allows for greater flexibility in teaching with the aim of promoting critical thinking and "learning how to learn." This initiative started some time ago in 1998, has been re-emphasized in the 2008 curriculum revisions. This established program allows local schools to actively engage in students' science learning and provides opportunities to stimulate science learning through an interdisciplinary approach. Similarly, the long-established lesson study approach underlines teacher agency in improving science teaching in the classroom.

Future directions for science education in Hong Kong include the expectant incorporation of information and communication technology (ICT) in science education and the feature of environmental education in the curriculum. The role of science, technology, engineering, and mathematics (STEM) has also recently been advocated in line with policy reorientations in science education. The Hong Kong government's 2015 Policy Address has underlined the role of the Education Bureau in renewing and enriching the STEM curricula. Multi-stakeholder initiatives have since taken place in support of this call. However, with evident partisan agendas, effectiveness of such activities is questioned. The STEM$^+$ program in China, which focuses on STEM Plus (+) humanities and arts, was launched in 2014 in selected primary schools in Shanghai to promote interdisciplinary abilities among students. While this program aims to veer away from traditional teaching, problems arising such as lack of teachers and too much reliance on executive decisions are apparent. Korea's revised science curriculum also targets to promote interdisciplinary thinking where science inquiry is highlighted through critical and creative problem-solving skills. Similarly in Taiwan, the new curriculum guideline which will be implemented in 2018 aims for the "cultivation of scientific literacy."

The assessment culture in East Asian countries (Kell and Kell 2013) clearly reflects and influences science education policy alignments among the country case studies discussed in this book. For instance, the National Assessment of Academic Ability (NAAA) started in 2010 in Japan, with the inclusion of assessment in science from 2012 aims among others to assess the academic achievement of students nationwide and to broadly improve education and educational outcomes. While Taiwan has earlier introduced a similar study in 2005, the Taiwan Assessment of Student Achievement or TASA, which intends to collect educational information, is broadly defined, from primary, middle, and secondary schools in Taiwan. This assessment practice not only allows a nationwide "standardized" analyses of science achievement in schools but is also considered helpful in gauging Taiwan's science student achievements with other countries. Additionally, it is anticipated that by 2018, primary teachers in Taiwan will be required to have a "science expertise" accreditation.

But unlike Japan and Taiwan, Korea's National Assessment for Educational Achievement (NAEA) was abolished in 2013 due to criticisms that underscore how NAEA mainly adapts teaching in the sciences and other subjects, for the test. The intense competition among schools is another reason for dismantling the nationwide assessment practice. Notwithstanding criticisms of national and standardized assessment practices (see Black and Wiliam 1998), Singapore is introducing a steady shift towards a "more creative or novel ways of asking questions" in the primary school leaving exam (PSLE) which will hopefully improve teaching in the sciences with "lessening the culture of performativity."

11.6 Closing Remarks

Looking at the present status of primary science education discussed in this book, the path looks bright in that the science curricula, pedagogies, teacher education, and recent reforms are on track with global needs and changes. In all the countries, assessment is an important component, and the present trend is to compare the status of science achievement between nations especially through TIMSS and PISA. The various descriptions of the metamorphosis of primary science curricula in the six nations over the past six decades reveal that the focus at present is upon inculcating values, attitudes, and skills in general and in particular to the environment. Much of the broader primary science curriculum in the past decade has been influenced by the United Nations Decade for Education for Sustainable Development (U.N. DESD) (Laurie et al. 2016), which exemplifies such importance:

> . . .to build young people's scientific knowledge; to increase their learning capacity; to impart upon them the values and lifestyles required for sustainable development. . .. (Gendong 2010: p.2 quoted in CS, China in UN DESD Report 2014)

> Activities involving museums and local communities are good opportunities for students to better understand local society and environments. ESD activities involving local community role-players, led to reciprocity by helping improve partnerships between schools and local societies. (Laurie et al. 2016, p.238)

Such revision of science content in the curricula has allowed children to enjoy science lessons, carry out hands-on science, and go on field trips while acquiring scientific skills through active learning. Additionally, interdisciplinary teaching and learning of science through STEM approaches should also be lauded as a step in the right direction. However, the planting of seeds of curiosity in the young mind (Wellcome Trust 2014) is only half the battle won. Pause and consider the following questions once again: "Why is primary science education important?" Despite all the engagement and activity in science classrooms, can our primary school children answer questions such as: "What is the nature of Science?," "What is a rainbow?," "Why does a river flow down a mountain?," or "Why are turtles in danger after surviving millions of years?" Why should pedagogies in science teaching continue to evolve and become more creative and relevant? How else can science teachers be trained to help students face a growing uncertain future? The answer perhaps lies in the reality that the children of today are all growing into future global citizens who will have to work together to continue to solve global problems that are threatening the world. Science teaches children from a young age about communication with others and about the world around them and helps ignite thinking minds to solve problems (Thomson 2017). Continuous monitoring and research must be done simultaneously to determine what needs to be further enhanced and changed in the curricula, pedagogies, teacher education, and reforms in the subject.

The main concern being: How can primary science education in the six nations leap to the next level? What can be a snapshot of the future in primary science education? Perhaps technologies such as virtual and augmented reality could further enhance content knowledge and skills coupled with increased interest and motivation for learning the sciences. Assessment in science learning perhaps could be more "for learning" and "as learning" and less of "of learning." Another aspect to be considered in primary science education is that of future professions in the scientific field (Thomson 2017). Research in the medical sciences and environmental-related sciences such as renewable energy and conservation will continue to increase and change. Globalization as it progresses will guarantee that children from Asian countries, such as those discussed in this book, are no exception and will need to face these changes. Finally, it cannot be denied that the wider curriculum that has been described indicates that the DESD (2005–2014) of the UN was instrumental in giving direction to science education. It has overwhelmingly aligned primary science curricula in trying to bring about changes in behavior, attitudes, and values from a young age (UNESCO 2016; Eilks 2014). As globalization continues on its track, global agendas like the UN's 2030 Agenda for Sustainable Development will further give direction and drive to primary science education in the coming years.

References

Battro, A., Lena, P., Sanchez-Sorondo, M., & von Braun, J. (2017). *Children and sustainable development: Ecological education in a globalized world.* Cham: Springer International Publishing AG.

Black, P., & Wiliam, D. (1998). Assessment and classroom learning. *Assessment in Education, 5*(1), 7–74.

Brooker, L. (2005). Learning to be a child: Cultural diversity and early years ideology. In N. Yelland (Ed.), *Critical issues in early childhood.* London: Open university press.

Caires, S., & Almeida, L. (2005). Teaching practice in initial teacher education: Its impact on student teachers' professional skills and development. *Journal of Education for Teaching, 31*(2), 111–120.

Cannon, J. (1997). *Influence of an extended elementary science teaching practicum experience upon preservice elementary teachers' science self-efficacy*. Available at: https://www.researchgate.net/profile/William_Mccomas/publication/262484186_Enhancing_undergraduate_science_instruction_through_science_education_partnerships_The_G-STEP_model/links/00463537c9cbb623ae000000.pdf#page=259. Accessed 2 Mar 2017.

Carter, L. (2005). Globalisation and science education: Rethinking science education reforms. *Journal of Research in Science Teaching, 42*(5), 561–580.

Childs, P. E. (2015). Curriculum development in science – Past, present and future. *LUMAT, 3(3)*, 381–400.

Darling-Hammond, L. (2000). How teacher education matters. *Journal of Teacher Education, 51*(3), 166–173.

Eilks, J. (2014). Science education and education for sustainable development – Justifications, models, practices and perspectives. *Eurasia Journal of Mathematics, Science & Technology Education, 11*(1), 149–158.

Gendong, S. (ed). (2010). Advance the construction of high-quality schools in basic education by education for sustainable development. [Project report]. Beijing, Chinese National Working Committee for ESD.

Goodlad, J. I. (1990). Studying the education of educators: From conception to findings. *Phi Delta Kappan, 71*, 698–701.

Huang, I. (2012). Science education in Asia. *Frontiers, 3*(02), 1–6.

Kell, M., & Kell, P. (2013). *Literacy and Language in East Asia*. Singapore: Springer.
Langford, R. (2010). http://journals.sagepub.com/doi/pdf/10.2304/ciec.2010.11.1.113
Laurie, R., Nonoyama-Tarumi, Y., Mckeown, R., & Hopkins, C. (2016). Contributions of education for sustainable development (ESD) to quality education. *Journal of Education for Sustainable Development, 10*(2), 226–242.
Loughran, J., Friedrichsen, P., & Berry, A. (2015). *Re-examining pedagogical content knowledge in science education*. New York: Routledge.
Magnusson, S., Krajick, J., & Borko, H. (1999). Nature, sources, and development of pedagogical content knowledge for science teachers. In J. Newsome & N. Lederman (Eds.), *Examining pedagogical content knowledge* (pp. 95–132). Dordrecht: Springer.
McNally, J., Cope, P., Inglis, B., & Stronach, I. (1997). The student teacher in school: Conditions for development. *Teaching and Teacher Education, 13*(5), 485–498.
National Inst. for Educational Research, Tokyo. (1999). *An International Comparative Study of School Curriculum. A Research Report* (ERIC Document, ED437711).
New Zealand Council for Educational Research (NZCER). (2010). *Primary science education for the 21st century: How, what, why*. Available at: http://www.nzcer.org.nz/system/files/primary-science-education-21st.pdf. Accessed 2 Mar 2017.
Thomson, C. (2017). *Why is science important in young kids' lives?*. Australian Institute of Policy and Science, New South Wales. Available at: http://www.schoolatoz.nsw.edu.au/homework-and-study/other-subjects-and-projects/science/why-science-is-important-in-young-kids-lives. Accessed 1 Apr 2017.
UNDESD Report. (2014). *Shaping the education of tomorrow*. Paris: UNESCO.
UNESCO. (2016). *Science for a sustainable future, In ESD success stories*. Available at: http://en.unesco.org/themes/science-sustainable-future. Accessed 1 Apr 2017.
Van Eijck, M., & Roth, M. (2012). *Imagination of science in education*. Dordrecht: Springer.
Wellcome Trust. (2014). *Primary science: Is it missing out? Recommendations for reviving primary science*. London: Research Report UK.
Wong, S. L., Wan, Z. H., & Cheng, K. L. (2014). One country, two systems: Nature of science education in mainland China and Hong Kong. In M. Matthews (Ed.), *International handbook of research in history, philosophy and science teaching* (pp. 2149–2175). Dordrecht: Springer.
Zhang, W., & Birdshall, S. (2016). Analysing early childhood educators' science pedagogy through the lens of a pedagogical content knowing framework. *Australasian Journal of Early Childhood, 41*(2), 50–58.
Zhang Lai Miao Li. (2000). *History of science education in the Republic of China history of science education in Taiwan, ROC*. National Academy for educational research, Taiwan. Available at: http://terms.naer.edu.tw/detail/1302475/. Accessed 2 Mar 2017.

Chapter 12
The Role of Teacher Education in Advancing Reform in Primary Science Education

Carla Zembal-Saul

12.1 Introduction

We are living in a world where rapid advances in science and technology are impacting society, the environment, and international interactions continuously. As a result, there is increasing recognition of the importance of school science and STEM (Science–Technology–Engineering–Mathematics) education to personal growth and career development, as well as productive participation in local and global society. This context frames, in part, some of the stated goals of primary science education of the East Asian states and territories featured in this volume. These goals are balanced against a collection of related concerns about the current state of primary science education that are consistent with those in the United States and numerous other countries. In the face of obstacles, innovation emerges. The authors of these reports provide thoughtful considerations, insights, and advancements in curriculum, instruction, and assessment from which we all can learn.

It is relevant to note my commentary is crafted against the backdrop of a *new* vision for students' meaningful science learning that has emerged in the United States in recent years. This vision of science learning is based on contemporary research and reflected in the *Framework for K-12 Science Education* (NRC 2012) and the *Next Generation Science Standards* (NGSS Lead States 2013). These ambitious standards have further highlighted gaps in the current state of science teaching and learning at all grades and the vision for students' science learning. Throughout the chapter, I will refer to these reform documents and the ways in which they are being addressed via research and practice to maximize the potential for productive implementation.

C. Zembal-Saul (✉)
The Pennsylvania State University, University Park, PA, United States
e-mail: czem@psu.edu

Y.-J. Lee, J. Tan (eds.), *Primary Science Education in East Asia*, Contemporary Trends and Issues in Science Education 47,
https://doi.org/10.1007/978-3-319-97167-4_12

With regard to the country reports, I will first address three problems of practice associated with the current state of primary science education shared across contributing countries – (1) student performance on international comparison tests, (2) the lack of hands-on investigation in primary science, and (3) primary teacher preparation in science. Then I will focus my comments on the challenges and promise of primary teacher education across the professional continuum (preservice to in-service), insights from research and practice focused on improving opportunities for teacher learning, and exemplar approaches in teacher education from contributing countries that are already underway.

As I write this chapter, I am keenly aware that although I was born and lived in Hong Kong and have traveled to Japan and Taiwan, I cannot know the lived cultural nuances of the educational and sociopolitical contexts of the East Asian states and territories featured here. Instead, I write from my experience in research and practice associated with preservice and in-service elementary teacher education in science across my career of 20+ years. I am motivated by our shared vision of the importance of primary science education, the need to make sense of persistent problems of practice, and the opportunity to learn from one another as a community of scholars and educators.

12.2 A New Vision for K-12 Science Education

In 2013, the *Next Generation Science Standards* (NGSS Lead States 2013) were released in the United States, replacing the *National Science Education Standards* (NRC 1996). The new and ambitious vision for K-12 science education states that, "[by] the end of 12th grade, *all* students [will] have some appreciation of the beauty and wonder of science; possess sufficient knowledge of science and engineering to engage in public discussions on related issues; are careful consumers of scientific and technological information related to their everyday lives; are able to continue to learn about science outside school; and have the skills to enter careers of their choice, including (but not limited to) careers in science, engineering, and technology" (NRC 2012, p. 1).

The NGSS were not released in isolation, but rather built on a steady succession of research-based reform documents, including the research synthesis, *Taking Science to School* (NRC 2007) and the *Framework for K-12 Science Ed*ucation (NRC 2012). Since the release of NGSS, additional reform documents that address assessment (NASEM 2017; NRC 2014) and science teachers' learning needs (NASEM 2015), which are necessary for successful implementation, have followed. The standards themselves are written as learning performances that pair content objectives with scientific and engineering practices. The ways in which big ideas are introduced and become more sophisticated over time are grounded in learning progression research (NRC 2007). This line of research coupled with findings that suggest even young children are capable of complex scientific reasoning has influenced the inclusion of science standards in Grades K-5.

Although it may be enticing to focus sole attention on the NGSS, one should read and become familiar with the *Framework* in order to better understand the underlying shifts represented in the new standards. The NGSS vision of students' science learning is frequently referred to as three dimensional (3D). In this context, 3D is intended to represent the interconnection of disciplinary core ideas (DCIs), scientific and engineering practices (SEP), and crosscutting concepts (CCC). DCIs are what we have traditionally thought of as the science content to be learned and used – life science, earth and space science, and physical sciences. In addition, there are standards for engineering and technology. Scientific and engineering practices are distinct from what has traditionally been taught in the United States as "the scientific method." They are intended to reflect three spheres of activities of scientists and engineers – investigating phenomena, arguing from evidence and evaluating claims, and constructing, using, and revising explanations and solutions (NRC 2012). Research has shown that learning DCIs while simultaneously engaging in SEPs enhances students' understanding of both science concepts and how knowledge in science is socially constructed (NRC 2007). Crosscutting concepts, which have previously taken a backseat to other aspects of science learning, are featured prominently in the NGSS. CCCs include structure and function, cause and effect, systems and system models, and other powerful lenses for interrelating knowledge across science domains to provide a coherent scientific world view.

As the full gravity of NGSS-inspired shifts in curriculum, instruction, and assessment sink in, it is evident that there is a significant gap in the United States between the new vision of students' science learning and actual practices in school science. This is especially true for our most vulnerable, high-poverty populations (NASEM 2015). Readers should note that at the elementary level, the state of science education in the United States is similar to what is described in the East Asian reports.

12.3 The State of Primary Science Education

As described in the country reports, primary science education in East Asia typically begins in Grade 3. Grades 1–2 often take an integrated approach, with emphasis in early literacy and numeracy. Although there are pockets of innovation and excellence, most authors here acknowledge an awareness that access to high-quality science education in primary grades is uneven and generally uncommon. That is, there are insufficient opportunities for students to engage in hands-on investigations of natural phenomena that involve productive participation in scientific discourse and practices. The report from China suggests real differences in science education opportunities between urban and rural schools, for example. The Singapore report explains that there is a great deal of societal pressure to do well in school and those who have the money to do so invest in private tutoring for their children. Thus, it is not surprising that so many East Asian countries embrace the notion of educational equity. Take, for instance, Taiwan, in which the first National Science Education Conference in 2002 produced a white paper that addressed their vision – science education for all citizens – marking a shift from science for the elite to science for all.

In looking across the reports featured in this volume, I identified several common problems of practice, which are listed below:

- Students' performance on internationally comparison tests, such as TIMSS and PISA, is not reflected in their attitudes and interests in the sciences.
- High-stakes testing is driving teachers' instruction to focus on textbooks and workbooks at the expense of hands-on investigations and inquiry.
- Primary teachers have limited backgrounds in science and scientific inquiry. They often lack confidence in their ability to teach science effectively and/or resist teaching science.

Each problem of practice is discussed in an attempt to illustrate the important role of strong alignment between the emerging vision for children's science learning, standards that include scientific practices, curriculum materials and resources, and consequential instruction. This kind of alignment also relies on coherent systems of teacher learning and ongoing development in science. It is a focus on teacher preparation and professional development that I will contribute to the conversation of primary science education in East Asia.

12.3.1 Student Attitudes and Interest in Science

It is striking that countries whose primary students perform so well on international tests should concurrently score so low on questions of attitudes and interest in science. In an age where science and technology is advancing rapidly, and associated careers are seen as essential to economic growth and independence, this phenomenon, and primary science education writ large, may even be characterized as being in crisis (Wu et al., Chap. 4, this volume).

Students' lack of interest in science and technology, however, is not an isolated problem. Jenkins and Pell (2006) published results of The Relevance of Science Education (ROSE) Project in England. The ROSE Project is part of a larger international comparison study of school science. It is distinct from other international comparisons because it asks students themselves about the relevance of school science education rather than placing emphasis on evaluating cognitive domains. Among the findings from England, it was reported that in addition to some notable gender differences, students believed science and technology to be important to society. Yet, their positive views of science and technology were divorced from their school science experience. Only a minority of students found science interesting and indicated that they liked it better than other subjects. These findings are fairly consistent with other countries participating in the project.

The disconnect between students' performance in science and their attitudes has prompted a number of East Asian states to focus specific attention on engaging students in investigations that are connected to their interests, solving problems that are relevant to daily life, and targeting scientific literacy for all. For instance, Taiwan is using findings of international comparisons to support the recommendation that instruction needs to be more learner-centered and less focused solely on cognitive domains (Wu et al., Chap. 4, this volume).

A potential asset of the states and territories represented in this volume is that each has a strong national curriculum that has undergone cycles of revision in order to reflect the findings of contemporary research and rapid advances in science and technology in society. Most East Asian countries identify inquiry and scientific literacy as cornerstones of their vision for science education at all levels. Educational reforms in science education have shifted from a knowledge focus to one that includes inquiry and scientific literacy. Nevertheless, authors have reported that the amount of time in the primary curriculum devoted to science is insufficient in comparison to the emphasis on mathematics and language learning.

Great effort has been taken to ensure that curriculum resources and textbooks have been closely aligned with national goals and standards. This kind of alignment is seen as fundamental to educational reform (NRC 2012). China reports on constant revision of their science curriculum and instruction, with strong recognition of the need to coordinate policy, curriculum, teaching, teacher education, and educational research. Rightly, Kim et al. (Korea, Chap. 5, this volume) assert that the curriculum is more than merely the textbook. In the small island state of Singapore, where the national curriculum is revised every 6 years, teachers are encouraged to adopt more student-centered approaches *through* the curriculum and associated resources. Lee (Chap. 6, this volume) describes the comprehensive national curriculum, which includes robust supplemental materials for teachers, such as lesson plans, activities, and workbooks. Nevertheless, most countries report limited enactment of investigation-based science at the primary level, which likely contributes to students' negative attitudes and lack of interest in science. This is also true in the United States where elementary school science is often not taught at all in early grades. When it is included in the curriculum, the focus is on either reading *about* science (i.e., limited/no investigations) or hands-on activities that do not connect with robust science learning objectives (e.g., "activity-mania").

The lack of hands-on investigations in primary classrooms is frequently attributed, at least in part, to an overreliance of teachers on textbooks and workbooks. In particular, concerns are raised in the report from Korea that textbooks do not allow for teacher creativity and independent decision-making. This situation is not unique to East Asian states and territories and is but an indication of other deeply seated issues related to the preparation of teachers and supports for their ongoing professional learning, which will be taken up later in this chapter.

12.3.2 Impacts of High-Stakes Testing

A related factor that country reports acknowledged here as contributing to a text-based, cognitive emphasis in primary science is high-stakes tests used for placement in secondary schools and/or other national and international assessments. While it is increasingly accepted that coherence among standards, curriculum, instruction, and assessment is necessary to make progress toward productive system-wide educational change (NRC 2012), there is awareness that these tests focus on science content with little, if any, attention to inquiry. In an atmosphere of high-stakes testing that focuses on content, it is not surprising that teachers resort to instruction that reflects this emphasis.

Recent advances in research on young people's learning in science have been synthesized in *Taking Science to School* (NRC 2007). Findings consistently assert that children come to primary school with the capabilities to participate productively in scientific discourse, practices, and reasoning; however, curriculum, instruction, and assessment systems are not typically in place to support their meaningful learning in science. Studies highlighted in this report are clear that when students engage in scientific practices simultaneously with learning science concepts, they learn both the science and about how science is done more deeply. In addition, leveraging children's natural curiosity about how the world works has been shown to be motivating and engaging for them.

It stands to reason that as more emphasis is placed on teaching to the text/test, which admittedly does not include many opportunities for inquiry or first-hand investigation, students would express low interest and unfavorable attitudes in science. The negative impacts of high-stakes assessments have influenced policy-makers to increase attention to students' attitudes as part of the curriculum revision process. In Japan, for example, there is new emphasis on fostering enthusiasm for learning and learning how to learn, as well as the application of knowledge to real-world situations.

The important role of assessment is acknowledged by the countries in this volume. Many are making important shifts to include artifacts beyond paper and pencil tests. For example, in Singapore they are moving to more application questions in science, which is confronting the "assessment bottleneck." In Japan, teachers are encouraged to use a variety of artifacts from instruction as a means of evaluating primary students' growth and development. In Korea, there are some provinces that are moving toward performance assessments as opposed to traditional examinations. In classrooms, teachers are able to determine the specifics of their assessment plans. There are websites and blogs through which teachers share lesson plans and test items. Likewise, in Taiwan, primary teachers are encouraged to use formative assessments and portfolios. However, it is often the case that these approaches have been found to be time-consuming and unreliable, so few teachers participate.

Concerns over high-stakes testing are not unique to East Asia. There is growing recognition that pressing for assessment reform in science education may serve as a powerful lever for change. There has been significant progress in China on the use of performance-based assessments in addition to high-stakes assessments. Hu and Shou (Chap. 3, this volume) provide examples of inquiry-focused assessments in their chapter.

In the United States, post-NGSS reports on assessment emphasize the importance of formative, classroom-based assessments that are aligned with standards, curriculum, and instruction. A seminal conclusion of *Developing Assessments for the Next Generation Science Standards* (NRC 2014) is that a coherent, multilevel assessment system is necessary for successful implementation of NGSS. Such a system would include classroom assessment used as part of everyday science instruction, assessments for monitoring student performance over time, and indicators for monitoring access to high-quality instruction for every child.

12.3.3 Perennial Issues with Primary Science Instruction

There are persistent issues associated with science education at the primary level that are shared across international boundaries, East and West. In addition to those mentioned previously, other barriers such as lack of materials for conducting investigations, or in the case of Singapore, lack of specialized staff to deal with equipment for science experiments, are well documented.

Shared concerns about the quality of primary science education include the preparation of teachers to support young children's science learning. More specifically, these concerns include teachers' lack of understanding and experience with science content and scientific inquiry, which likely contributes to their lack of confidence with teaching science to children. In addition, the Hong Kong report points out that not only do teachers experience difficulty with the science content itself but also in assisting students who struggle with particular content. This reflects limitations in specialized teacher knowledge necessary for supporting the learning of others, known as pedagogical content knowledge (PCK) (Grossman 1990; Magnusson et al. 1999; Shulman 1986). As addressed in a number of country reports, the preparation of primary teachers as generalists is a key limitation to high-quality science instruction.

In the remainder of this chapter, I will focus on the importance of coherent and ongoing education experiences for primary teachers in science as one inroad to reform. To be clear, it will require a constellation of coherent initiatives, which include assessments, administrative support, and district and school policies and practices, to make significant progress. There is much to be learned about advances in teacher education from the approaches of the countries represented in this volume.

12.4 Toward a Coherent System of Teacher Education in Science

In the past, I have written extensively about the importance of providing beginning and practicing teachers with a coherent framework for supporting children's learning in science (Zembal-Saul 2016; 2009). Research that examines preservice teachers' learning and development within the context of programs informed by explicit conceptual frameworks report that their understandings and practices ultimately reflect central components of those frameworks (Clift and Brady 2005). Darling-Hammond et al. (2005) suggest that such frameworks can assist teachers in organizing their thinking and support learning throughout their careers. In terms of coherence, the authors state, "Repeated experiences with a set of conceptual ideas, along with repeated opportunities to practice skills and modes of analysis, support deeper learning and the development of expertise" (p. 393). Consider here that coherence includes teacher candidates' science learning experiences, methods of teaching science, and field experiences hosted by capable mentor teachers who understand and support the goals and vision for primary science education and are involved in ongoing professional development themselves.

As science education increasingly focuses on learning science concepts through engaging in scientific discourse and practices, such as arguing from evidence and constructing and interrogating scientific explanations about real-world phenomena, the stakes for primary teachers increase. Not only is it unlikely that they have been prepared to understand the science content they will encounter as primary school teachers, but they also are unlikely to have experienced learning in science that reflects giving priority to evidence and explanation – concerns echoed in Taiwan's report. This reality requires that we not blame primary teachers for their lack of preparation in science and science teaching. Rather we should consider the kinds of coherent learning experiences that stand to be productive for preservice and in-service teachers across the professional continuum as they go about supporting this new vision of students' science learning in their classrooms.

12.4.1 Primary Teacher Preparation

There are interesting overlaps, as well as variation, in how the East Asian states and territories in this volume approach primary teacher preparation. Some, such as Singapore, point to advances in teacher preparation as having a positive impact on students' performance on high-stakes national and international comparison tests. A consistent shift in thinking about teaching as a profession, in contrast to a vocation, is relatively recent and underlies reform initiatives. For example, in China this shift to teaching as a profession occurred around 2000. They now have more than 60 institutions approved by the Ministry of Education with science education departments, as well as multiple pathways for primary teachers to be prepared to teach science. These pathways range from traditional teacher training colleges to undergraduate university programs that prepare primary teachers as science specialists.

Although there are emerging pathways for specialization in science education, most primary teachers do not choose to pursue them. As a result, they often encounter difficulty when attempting to teach science. For example, in Hong Kong five universities offer Bachelor of Education programs that prepare teachers for the primary curriculum. These programs include a course on science and technology, teaching methods, and a supervised research project. Most preservice teachers elect to focus on their major, which is pedagogy; however, a few select other subjects including science. The report from Korea describes teaching as a popular career choice that is stable and well respected. In light of this, there is an oversupply of teachers and competition for entrance into teacher education programs. Primary education is offered at designated universities that focus solely on teacher education. Programs follow a 4-year cohort model in which candidates learn to teach all subjects. Only a small number of total credits focus on primary science, however. In Taiwan, the Teacher Education Act of 1994 allows all universities to provide teacher education programs. Yet, there are no dedicated departments for primary teacher preparation, and preservice teachers are not required to take science or science teaching methods courses.

The Singapore report describes an enviable system of teacher preparation in terms of potential coherence. Most teachers pass through the National Institute of Education, which is their sole teacher training college. Preservice teachers complete a 4-year undergraduate degree with field experiences and community service. A postgraduate program is offered and intended to strengthen theory–practice connections and also includes field experiences. Coherence is built into this system given that field experiences are necessarily linked to the classrooms of teachers who have been through the program. This increases the potential that new teachers will see in practice what they are learning about at university and that in-service teachers will be introduced to new ideas and methods advanced through university studies. Unfortunately, as with other countries, few primary teachers pursue science majors.

At my institution, I have had the opportunity to study the longitudinal development of undergraduate elementary teacher education majors. Beginning even before they enter the major, my research group has examined the influence of collaboratively designed science content courses for undergraduate, non-science majors preparing to be teachers. For example, a course that uses insects as models of biological evolution was designed and taught by science education and entomology faculty and graduate students for future primary teachers (Haefner et al. 2006; Haefner and Zembal-Saul 2004). This course provided opportunities for education majors to learn relevant science content by working in small groups, engaging in scientific discourse and practices, and considering applications to teaching science to children. Undergraduates who completed the course demonstrated an increased understanding of scientific inquiry and science content. In addition, they reported less anxiety about teaching science when subsequently taking their elementary science methods course. We now have a repertoire of specialized life, earth and space, and physical science options for education majors preparing to be primary teachers.

The final year of teacher preparation for the future teachers I study is a field-based professional development school (PDS) partnership between the College of Education and the local school district (Badiali et al. 2012). Future teachers spend an entire year in schools with mentor teachers, and methods courses are co-taught by university faculty and experienced teachers. Video analysis of practice among preservice teachers and mentors is commonplace and accepted as something that effective teachers do. A hallmark of the program is that preservice teachers and some mentors participate in practitioner inquiry (Badiali et al. 2012). That is, they identify and systematically examine problems of practice from the classroom. The results of this "action research" are made public at an annual Inquiry Conference attended by hundreds of teachers and community members who learn from one another. This model of teacher preparation is built on a foundation of simultaneous renewal, and it aims to enhance coherence between university coursework and classroom practice.

12.4.2 Professional Learning Opportunities for Teachers in Science

In 2015, the US National Academies of Sciences, Engineering, and Medicine (NASEM) published a report on science teachers' learning needs. In this report, the committee encourages educators, the public, and policy-makers to consider teacher learning as dynamic, collective, and mediated by context. A main idea is that we look across the professional continuum – from preservice to in-service education – and emphasize collaborative systems of innovation in science teaching. In addition, the report describes a consensus view about features of effective teacher professional learning opportunities in science that is grounded in research. These features include (1) content-rich opportunities for teachers to learn science and how children learn science; (2) active learning (numerous forms) paired with interactive feedback and discussion; (3) coherence with other learning experiences, as well as district policies and practices; (4) over time and with sufficient duration for cycles of implementation and reflection; (5) and capacity building through participation of groups of teachers from the same school/district. Moreover, the committee recommended that teacher professional learning opportunities be informed by the best available research and be both embedded in the school day and offered outside of school. Exploring the use of new formats and digital tools for professional learning was also encouraged.

The East Asian states and territory reports identify teacher education as a main focus for improving the quality of primary science education. This goal is addressed in teacher professional development through a variety of ways, many of which are aligned with the aforementioned NASEM (2015) report. Common threads include science-specific learning opportunities, online communities, teacher inquiry, and collaboration and sharing among educators, science specialists, and master teacher mentors in science.

In Taiwan, they have embraced the notion of job-embedded professional learning with their compulsory Wednesday afternoon professional development for teachers. Unfortunately, the focus of this time is typically on mathematics and language instruction, with few science learning opportunities provided. Although inquiry-based instruction is not common, teachers who have adopted this approach to teaching science are encouraged to share their lesson plans and other resources with colleagues. There are strong mentoring groups for primary science teaching that support instruction and provide science workshops. Local innovation feeds into a national mentor network, as well as other online teacher communities.

Given that most primary teachers are not science majors, Singapore's National Institute of Education provides multiple pathways for primary teachers to pursue advancement in science, including a postgraduate certificate and degree. There also are a large variety of professional development courses in STEM education, argumentation, problem-based learning, and more. The Academy of Singapore Teachers has a primary science chapter led by master teachers who provide professional development for classroom pedagogy. Teachers develop their own personal learning plans, and each receives 100 h of paid professional development. In addition, the Singapore Ministry of Education is "ramping up" the number of science specialists. Specialists provide just-in-time assistance with questions related to implementation of the national curriculum.

In Hong Kong, there is an understanding that continuous professional development for primary teachers is needed to raise their confidence with science instruction. They have 3-year cycles of professional development that require 150 h of learning, over half of which is focused on pedagogy. The Education Bureau provides and funds professional development workshops to meet the needs of teachers. They are, however, described as short term, and teachers do not have the opportunity to try out what they are learning in their classrooms. The Hong Kong Institute of Higher Education provides more in-depth workshops (1–2 days) that emphasize basic science and hands-on investigations. There also is promising research on platforms for exchanging resources and ideas, as well as video analysis of practice.

Approaches by which teachers systematically examine their own practice, such as lesson study and teacher research, also known as teacher inquiry or action research, portray teaching as a dynamic profession in which we continue to learn from and about practice. This kind of job-embedded approach to teacher learning is reflected in several country reports and holds much promise for positive change in science instruction. For example, in Japan, teacher learning communities are accepted and valued. Collaborative teacher inquiry via lesson study is a normative approach for teachers to investigate instructional practices and publicly share findings with peers and new teachers.

It is clear from the contributions in this volume that East and West are grappling with similar concerns about the quality of primary school science and the importance of teacher education in science. While context is a significant piece of the puzzle, there are ideas that transcend international boundaries. Moving forward, it will be essential that we view teacher learning as ongoing; that teacher learning opportunities are coherent and reflect the new vision for students' science learning; that teachers have opportunities to participate productively in science learning and scientific inquiry, as well as associated pedagogy; and that there are collaborative communities that continue to examine science teaching and assessment practices, as well as share resources, findings, and supports. Taken together, these are components of the dynamic and collective expertise of professional educators.

12.5 Final Remarks

I have greatly appreciated the opportunity to review the country reports in this volume on primary science teacher education in East Asia. We share much common ground when it comes to our goals for preparing a scientifically literate citizenry and workforce, which balances personal growth and development with national security, economic prosperity, and environmental well-being. Moving forward, it is tempting to embrace new standards as a panacea for improving science education, especially at the primary level. In order to learn from failed past reforms, however, we must consider the underlying research and framework, which calls for targeted shifts in the ways that science is learned. Additionally, we need to address complex systems of science education and strive to create more coherence among curriculum, instruction, assessment, and policy. The preparation and professional development of primary teachers to support children's meaningful science learning and scientific inquiry are critical components of a comprehensive, multipronged approach to science education reform.

As described in this chapter, the East Asian state and territory reports include a variety of initiatives that are aligned to recommendations for supporting teachers' learning needs in science (NASEM 2015). While coherent systems of teacher education are critical to support change in primary science education, it is important to note that such coherence can be thwarted when school, community, regional, and national policies and practices are not aligned. Although it seems productive to focus on the education of primary teachers across the professional continuum, if these teachers continue to work in settings that value and reward the results of high-stakes tests and do not foreground inquiry, positive impacts of reform through teacher education will be limited.

References

Badiali, B., Zembal-Saul, C., DeWitt, K., & Stoicovy, D. (2012). Shared inquiry: The professional development school as a laboratory of practice for preparing the next generation of teacher educators. In *Placing practitioner knowledge at the center of teacher education: Rethinking the policy and practice of the education doctorate*. Greenwich: Information Age Publishing.

Clift, R. T., & Brady, P. (2005). Research on methods courses and field experiences. In M. Cochran-Smith & K. Zeichner (Eds.), *Studying teacher education: The report of the AERA panel on research and teacher education* (pp. 309–424). Mahwah: Erlbaum.

Darling-Hammond, L., Hammerness, K., Grossman, P., Rust, F., & Shulman, L. (2005). The design of teacher education programs. In L. Darling-Hammond & J. Bransford (Eds.), *Preparing teachers for a changing world: What teachers should learn and be able to do*. San Francisco: Jossey-Bass.

Grossman, P. (1990). *The making of a teacher: Teacher knowledge and teacher education*. New York: Teachers College Press.

Haefner, L. A., & Zembal-Saul, C. (2004). Learning by doing? Prospective elementary teachers' developing understandings of scientific inquiry and science teaching and learning. *International Journal of Science Education, 26*(13), 1653–1674.

Haefner, L. A., Friedrichsen, P., & Zembal-Saul, C. (2006). Teaching with insects: An applied life science course for supporting prospective elementary teachers' scientific inquiry. *The American Biology Teacher, 68*(4), 254–259.

Jenkins, E. & Pell, R. (2006). *The relevance of science education (ROSE) project in England: A summary of findings*. Report published by the centre for studies in science and mathematics education. Leeds: University of Leeds.

Magnusson, S., Krajcik, J., & Borko, H. (1999). Nature, sources and development of pedagogical content knowledge for science teaching. In J. Guess-Newsome & N. Lederman (Eds.), *Examining pedagogical content knowledge: The construct and its implications for science education* (pp. 95–132). Dordrecht: Kluwer.

National Academies of Sciences, Engineering, and Medicine. (2015). *Science teachers' learning: Enhancing opportunities, creating supportive communities*. Committee on strengthening science education through a teacher learning continuum. Board on science education and teacher advisory council, Division of behavioral and social sciences and education. Washington, DC: The National Academies Press.

National Academies of Sciences, Engineering, and Medicine. (2017). *Seeing students learn science: Integrating assessment and instruction in the classroom*. Washington, DC: The National Academies Press.

National Research Council. (1996). *National science education standards*. Washington, DC: National Academies Press.

National Research Council. (2007). In R. A. Duschl, H. A. Schweingruber, & A. W. Shouse (Eds.)., Board on science education, division of behavioral and social sciences and education *Taking science to school: Learning and teaching science in grades K-8. Committee on science learning, kindergarten through eighth grade*. Washington, DC: The National Academies Press.

National Research Council. (2012). *A framework for K-12 science education: Practices, crosscutting concepts, and core ideas*. Committee on a conceptual framework for new K-12 science standards, board on science education, division of behavioral and social sciences and education. Washington, DC: The National Academies Press.

National Research Council. (2014). *Developing assessments for the next generation science standards*. Committee on Developing Assessments of Science Proficiency in K-12. Board on Testing and Assessment and Board on Science Education, J. W. Pellegrino, M. R. Wilson, J. A. Koenig, & A. S. Beatty (Eds.). Division of Behavioral and Social Sciences.

NGSS Lead States. (2013). *Next generation science standards: For states, by states*. Washington, DC: The National Academies Press.

Shulman, L. S. (1986). Those who understand: Knowledge growth in teaching. *Educational Researcher, 15*(2), 4–14.

Zembal-Saul, C. (2009). Learning to teach elementary school science as argument. *Science Education, 93*, 687–719.

Zembal-Saul, C. (2016). Implications of framing teacher development as identity construction for science teacher education research and practice. In L. Avraamidou (Ed.), *Studying science teacher identity: Theoretical, methodological, and empirical explorations* (pp. 321–332). Boston: Sense Publishers.

Chapter 13
Institutional Pressures on Primary Science and the New Ecosystem of Knowledge

Elizabeth H. McEneaney

In reviewing the chapters detailing primary science education, I was pleased to see a genuine focus on core issues of curriculum; that is, attention to what is taught and what is learned in this aspect of schooling. These system-level reports can therefore be seen as modest antidotes to what Michael Young has called "the crisis in curriculum theory" in that they "bring knowledge back in" to the focus of our scholarly attention (Young 2008, 2013; Baker 2014, 2015). The contours of primary school science in each system are shaped by historical roots in each country as well as global influences, such as ripple effects of the Sputnik shock of the late 1950s and international testing regimes. Reports in this volume indicate that primary school science is also influenced by the degree of centralization of the educational system as well as the organization of initial teacher training and ongoing professional development. Also exerting some influence are key research results that shed light on the content and structuring of the curriculum and the way it is delivered by teachers, though perhaps educational research on and about primary science in each of these systems exerts less influence on policy and practice than many of us would like.

In this chapter, I center my analysis on the primary science curriculum and, to a lesser degree, the pedagogy that implements it, and I consider the relatively long-standing effect of the broader institutional environment on what is taught and learned in these East Asian systems. I will also analyze what the rapidly changing technical environment might mean for the primary science curriculum. I argue that developments in information and communication technology (ICT) radically alter the "ecosystem of knowledge" in ways that curriculum makers must take into account explicitly if schooling is to empower pupils fully. Some educational systems are better positioned than others, in their guiding principles and particular

E. H. McEneaney (✉)
University of Massachusetts, Amherst, MA, USA
e-mail: emcenean@educ.umass.edu

Y.-J. Lee, J. Tan (eds.), *Primary Science Education in East Asia*, Contemporary Trends and Issues in Science Education 47,
https://doi.org/10.1007/978-3-319-97167-4_13

practices, to prepare pupils to leverage their access to information technology for powerful science learning.

In this way, I am drawing on organizational theory that posits two basic types of environments to which organizations respond, the technical environment and the institutional environment (Meyer and Scott 1983, Scott 2003). The technical environment is premised on the notion that organizations are rational and that actors can identify desired outcomes and the means to attain them, and it encompasses the technology (broadly construed as tools, machines, material resources, processes, and methods) as well as the technical expertise required to achieve outcomes. The institutional environment, in contrast, highlights regulative and normative processes that shape organizations but also cultural features such as schemas, categories, and discursive resources that can be used to justify organizational policies, such as national curricula, in an almost script-like manner. In terms of school systems, the state and accreditation agencies of various kinds exert regulative and normative pressure, while professional organizations of teachers, national and regional as well as international, are conduits of a variety of cultural pressures, often interpreted and packaged for application in schools by globally networked university-based education scholars. Although these two types of organizational environments are neither completely distinct nor opposing, they do pose separate demands on school systems, and we can see their effects on the development of curricula (and attendant pedagogy) over time. While the institutional environment pertinent to mass education has exhibited gradual but influential shifts over time, the technical environment, with the rapid rise and especially the expansion of access to ICT, is clearly in a state of substantial transformation. As such, what were fairly stable institutional arrangements will be challenged (see Scott et al. 2000 for an analysis of change in hospitals).

Though many readers of this volume might attend to the subtle differences described in primary school curriculum and pedagogy across the six systems, I was especially drawn to the remarkable similarities evident across these accounts regarding the aims and content of primary science (see also Lee et al. 2015) and the understandings about children as learners and prescriptions for how teachers should impart school science. These similarities are products of the relatively stable institutional environment and are manifest in their tendency to emphasize (a) interconnectedness and integrated themes, (b) scientific literacy, (c) learning-to-learn and skills for lifelong learning, (d) cultivation of values and attitudes, (e) constructivist learning, and (f) the centrality of inquiry and process skills. Correspondingly, across systems there is a notable absence of emphasis on specific scientific-technical knowledge, no comprehensive list of concepts and terms pupils should know. This strong trend toward a virtually "content-free" primary science (McEneaney and Meyer 2000) might be of concern to some, such as Michael Young (2013) who worries that lack of disciplinary grounding in the curriculum in science and other subject areas leaves too many young people without access to what he terms "powerful knowledge." In contrast, I will argue that curricula that wisely and explicitly leverage the internet and other ICT resources can allow in-school engagement with inquiry processes while also cultivating ties to science disciplines and disciplinary language and ideas.

In the next section, I highlight details of the many ways that similarities are manifested in the various system-level approaches to primary science described in chapters in this volume, while offering some explanation for this tendency toward isomorphism.

13.1 Institutionalist Views of School Curricula

Many sociologists working within a neo-institutionalist tradition have documented this strong global trend toward similarity or isomorphism in all sorts of activities sponsored by modern nation-states or entities aspiring to nation-statehood. The normative expectations for how to construct policy around these highly institutionalized activities, such as determining citizenship and making provisions for education and health care, are abundantly clear (Meyer et al. 1997) through what Boli and Thomas (1999) call a "world culture." This is true even as on-the-ground resources and practices may not live up to the idealized, global vision enshrined in policy, including official curricula and officially approved curricular materials like textbooks. These normative strictures have clearly shaped mass schooling and national curricular frameworks for many decades (Meyer et al. 2017) and now extend upward to what is taught and learned in tertiary education (Frank and Gabler 2006). They apply perhaps with even greater force to STEM fields at all levels, as building human capital in STEM has come to be understood as core to a nation's economic survival, even as there is not clear empirical evidence that this is true for all or even most countries (Schofer et al. 2000). From a neo-institutionalist perspective as well, schools are charged with constructing individuals imbued with agency, inclined toward action on a natural (and social world) that is deemed to be rationalized, ordered, and manageable (McEneaney and Meyer 2000, McEneaney 2003a). And it is the curriculum, as Popkewitz (1997: 139) notes, that regulates, in a Foucauldian sense, its young subjects, "produc[ing] sensitivities, dispositions and awarenesses." In this global model of the "schooled society," science becomes the main source of all truth claims, and it takes on a universalistic quality – the mantra is that scientific ways of knowing should be applied to ever-widening circles of human experience (Baker 2014, 2015) to the exclusion of other funds of knowledge, such as religion.

All of these forces come together to shape primary school science curriculum (and the pedagogy aiming to impart it) in especially profound ways. It is in the early grades that schools in modern systems embark in earnest on the project of constructing agentic individual persons and lay the groundwork for the idea that science thinking is justifiably applicable to all manner of problems in one's life. In saying this, I do not in any way wish to denigrate the importance of science education and the role that science literacy and the advance of science disciplines can play in improving the human condition. My point is the degree to which all of this has a taken-for-granted quality that manifests in turn in curriculum and pedagogy. As curriculum and pedagogical prescriptions resonate with these details of a "world culture," strong isomorphism results, and we can see it in the system-level descriptions in this volume.

Before outlining in more detail the level of isomorphism in East Asian primary school curricula evidenced in chapters in this book, it may be instructive to conduct a modest thought experiment (following in the spirit of Meyer et al. 1997). Try to imagine for a moment that a country elected a new leader who promised to bring disruptive change to all government institutions, including education. First, the new administration exerted substantial effort to portray science as a province of the elite, such that only a select few students would be expected to make a professional contribution to science fields in the future. New science textbooks and curricular materials therefore depicted "heroes" of disciplinary science (if they depicted people at all – administration officials voiced concern that this might distract from core content). Specialized versions of curricular materials were developed for subgroups seen as more likely to attain status as professional scientists (e.g., boys, majority races, students in urban areas). Classical problems of science were deemed most worthy of study, and soon classroom activities shifted toward a retracing of the established paths to solving these problems. Textbooks sequenced and structured the core disciplinary content in an authoritative manner. Teachers required students to memorize mathematical formulas, scientific constants, and periodic table facts as a way to honor the historical achievements of professional science. Science experts convinced the new administration to de-emphasize the everyday relevance to children of science in school curricula, while allowing depictions of its relevance at a societal level: its contributions to military superiority, economic development, preventing epidemics, and so on. They also warned against spending school time trying to engage the emotions or develop positive attitudes toward science learning and science more broadly because that "waters down" the curriculum. Funding was slashed for researching and assessing students' interest in, motivation for, or self-efficacy to do science. Approaches grounded in a "fixed mindset" soon came into vogue, along with assumptions that children destined to be professional scientists enjoyed the academic work and that they were inherently drawn to the aesthetics of disciplinary science and would naturally excel. For others, short-term mastery of the cognitive content became sufficient.

From a contemporary purview, such an approach seems virtually unthinkable in any national education system claiming to be modern in any sense. It strikes us that way not only as it might apply at the primary school level but also at higher levels of science education, even though such an approach is not so very different than the post-Sputnik curricula common in the United States as well as Western Europe and much of East Asia in the 1960s and 1970s. Since then, however, as the world cultural frame consolidated and has come to offer highly institutionalized notions of children as learners, science as a realm of human activity, and the purpose of schooling itself, a broad but very different consensus about school science has emerged. In short, this institutionalized consensus emphasizes integrated themes that cut across disciplinary boundaries, conceptions of scientific literacy including the notion that all students can and should learn and apply science thinking in their everyday lives (McEneaney 2003b), and cultivation of positive attitudes toward science. Constructivist understanding of learning is undisputed, and so primary science allots much time to inquiry activities and the development of process skills. Allowing pupils to work on problems of their own choosing, or at least problems that are relevant in their everyday lives, is understood to support the construction of meaning while enhancing interest and motivation in science and bolstering a belief in the general utility of science. This institutionalized consensus is evident in East Asian primary science curricula and related pedagogical prescriptions.

13.2 Similarities in Primary Science Across East Asia

Readers can easily find evidence of the robust, isomorphic pressures from the institutional environment in which national school systems operate in the various accounts of primary school science in this volume. These areas of thematic consensus have developed a taken-for-granted quality that marks their institutionalized status and, in the descriptions of the state of primary science here, address the nature of school science, who should learn and apply science, and how it is learned.

Interconnectedness and Integrated Themes From the perspective of national curricula, primary science in East Asia very typically is portrayed as deeply interconnected across spheres of life and across disciplinary ways of thinking. The historical trajectory of Hong Kong's curriculum from the early subject "Nature Study" to "Primary Science" to science as a core learning area (along with personal, social, and humanities education and technology education) in "General Studies" is typical of this integrated approach in contemporary primary science. It reflects a modern view of science as integrated and applicable to not only the natural world but also the social world and personal sphere as well. In its view of school science as involving the "study of phenomena and events around us," Hong Kong's curriculum grounds school science in the everyday. Taiwan's "life course" in grades 1 and 2 has some of these same features. Korea too now emphasizes both interconnected across disciplines as well as connections to everyday life, for example, by grouping science and home economics together in its 2009 reform.[1] Japan's notion of curricular "pillars" ultimately highlights science as part of a cross-disciplinary school knowledge base. For most of East Asia, this broad general studies approach narrows slightly heading into upper primary grades but still retains the near-requisite interdisciplinary thrust, often including technology, which insures relevant applications in everyday life. Thus, in breaking down disciplinary distinctions, while grounding science in the child's everyday world with the expectation that science be applicable in all facets of it, East Asian science curricula frame school science as integrated and, above all, relevant. This institutionalized perspective on the nature of school science is encapsulated by this comment about China's primary science "science comes from everyday life, teacher brings it back to life," although this sentiment

[1] Curiously, this relatively recent trend toward deeply interdisciplinary treatments of science in the primary school, with applications in the personal, then familial, community-based, and societal, harkens back to the older German sachunterricht tradition, which may have been before its time, institutionally speaking.

could emerge from science curriculum developers anywhere in East Asia or any country that has connections to the modern world polity.

Scientific Literacy All national systems represented in this volume take, as a given, the aim of promoting scientific literacy. In fact, only the most isolated, authoritarian nations in the world would eschew this as an aim for science education; though definitions vary (and perhaps to some extent because they vary), scientific literacy resonates with the institutional environment strongly and has long had a global cachet (McEneaney 2003b). Regarding the question of who should learn science, this approach in the first instance embraces a "science for all" message, widely observed in East Asian systems and embodied more practically in, for example, Japan's policy of mixing ability levels in science classes. Most East Asian systems described here (e.g., Japan, Hong Kong, and Taiwan) emphasize that pupils' learning of science should improve their capacity as citizens to make decisions. China frames its approach to scientific literacy differently, no doubt due to particular historical and political characteristics that slightly distance it from institutionalized "world culture." Nonetheless, the Chinese system as described here offers education as a "public right" with equal access "safeguarded," although it is likely that the right to science education in particular is inalienable in ways that the rights to boundless study of other school subjects may not be. Thus, this common understanding that primary science curricula (and in fact all school science) serves to further the goal of scientific literacy reflects an institutional environment that demands construction of individuals imbued with agency, ready to deploy science knowledge for both personal and societal betterment.

Learning-to-Learn and Skills for Lifelong Learning Another area in clear thematic consensus in the national system descriptions here is the emphasis on learning-to-learn and more generally the cultivation of skills for lifelong learning. This too is in keeping with institutional pressures toward individualistic agency, insuring that such dispositions endure through the life course and regardless of conditions. Learning-to-learn is an explicit aim in primary science in Korea, China, and Hong Kong, while in other systems (Singapore and Taiwan), the emphasis is on developing critical thinking skills. The Taiwanese curriculum aims to support "habits of reflection," while Japan's early integrated approach to science learning is intended to help pupils "think in their own way about life." Implicitly, both of these aim to produce habits and modes of thinking that are lifelong in duration, even as the subject of reflection might change over time.

Cultivation of Values and Attitudes In concert with scientific literacy aims and producing lifelong learners, primary science curricula in East Asia (as documented here) and in much of the rest of the world target the cultivation of values and attitudes. In Hong Kong's General Science curriculum, it was reported that learning activities aim to "arouse interest in… enquiring about science, technology and society," just as Taiwan's primary science encourages "participation in scientific discussions." In China, the goal is for students to see themselves as interested, motivated "investigators of science." These examples point to a sort of democratizing of science expertise; children too can be experts. Yet the affective elements of the institutional project assigned to mass schooling in institutionalized world society now go beyond constructing individuals who are interested in science. Japan's school science aims to help pupils develop a "zest for life" and to "nurture hearts and minds [with] affection for nature" in addition to fostering a scientific perspective. In other systems, curricula (and attendant pedagogy) aim to generate enthusiasm, joy, curiosity, and/or creativity. Some countries, like Korea, may worry that this set of goals, consistent with institutional pressures to not simply impart content but develop agentic persons with affective disposition to value science and act scientifically, is not being attained. Nonetheless, the fact that national curricula in primary science aim to develop these values and attitudes is evidence of the institutional environment's effects.

Constructivist Learning Regarding the questions of how pupils might best learn science, all systems take as absolutely foundational the idea of learning as constructivist. A core component of that thinking is a focus on creating child-centered curricula and instruction. We see it in Taiwan's reported move toward a more learner-centered primary science and Japan's pattern in the postwar period to value "creative learning… according to [children's] interests and needs." Institutionalized, discursive scripts seem to be enacted as when China embraces "learning by doing," "active learning," and "collaborative learning" and Singapore adopts "problem-based learning." Singapore's "knowledge building" pedagogy is a local application of a model developed in the West (Bereiter and Scardamalia 1996; Scardamalia and Bereiter 1996). All these examples are consonant with global institutional pressures, are highly constructivist in principle, and, in the case of the knowledge building approach, are providing supports for constructivist argumentation.

The Centrality of Inquiry and Process Skills Another thoroughly institutionalized element of primary science, as demonstrated in the detailed descriptions of national system approaches in this book, is the centrality of inquiry in the primary science curriculum. The US National Research Council's statement on inquiry offers a type of authoritative discursive resource that several systems (Singapore, Taiwan) cite to justify emphasizing inquiry. Taiwan views it as a "core characteristic of science education," while the report here from Singapore describes inquiry as "the defining feature, the heart of instruction" as teachers are trained to implement a constructivist, "collaborative inquiry." In all of this, there is an emphasis on critical thinking, as noted above, along with support for more structured inquiry and scientific skills and processes. Chinese primary science aims to engage pupils in "participating in the identification of evidence, making judgments, reasoning logically, and thinking about how to interpret data." Both Chinese and Singaporean systems are said to go beyond expectations for "hands on" engagement toward, as in Taiwan and elsewhere, "hands on, minds on." Recent Chinese reforms emphasize spirally arranged training in "learning to think," while Singapore stresses "evidence-based processes" along with hands on activity. While the reports acknowledge the varying degrees of difficulty teachers have in actually implementing this vision of school science, the consensus, enshrined in official curricular statements and policies, is that science inquiry occupies hallowed space in a modern science education at the primary level. What pupils are to be enquiring about, i.e., content as opposed to process, is left unspecified, except that it should be relevant to the child. Consistent with this radical child-centered position, now thoroughly institutionalized, is China's view of "inquiry-as-content." Importantly, for curriculum theorists, this raises the question of whether there is any requisite content in the primary science curriculum that is distinct from inquiry or process skills, and if so, what is it. Might there be specific facts or disciplinary topics, language, or terms that pupils would need to learn *prior* to embarking meaningfully on inquiry experiences? Or can it all be constructed or co-constructed – more meaningfully – through inquiry? I view Michael Young's (2013) concern about schools failing to provide access to "powerful knowledge" as related to this kind of uncertainty about the proper relative weight of content knowledge versus process skills in curricula for a range of school subjects, science included.

To summarize at this point, what might seem strange in all of these national system reports is the absence of strangeness. There is very substantial sameness across these East Asian cases in descriptions of interdisciplinarity, the explicit cultivation of values and attitudes, the prizing of scientific literacy and inquiry, and the bedrock belief in inquiry as a mode of learning. The absence of strangeness also lies in the fact that these reports could have emerged with very few differences from any number of places around the world, not just East Asia. Yet, institutional theory offers an explanation for these isomorphic tendencies, given a relative stasis in the technical environment related to mass education. In the next section, I argue that rapidly expanding access to ICT is the sort of game-changing element in the technical environment that may wrest influence from the institutional environment, with some national systems better positioned to respond than others.

13.3 Technical Conditions Shaping Twenty-First-Century School Curricula

Young's concern (2013) about the potential for students of all ages to access "powerful," discipline-based knowledge through school curricula, what he characterizes as "epistemic access" following Morrow (2008), comes at an interesting time. As I demonstrate above, much of the school curriculum (and its related pedagogical prescriptions) is thoroughly institutionalized. This is strikingly true in primary science, targeted at an age level where schools are charged with developing agentic individual, and in a field thought to be central to the development of modern nation-states. In contemporary primary school science, we see the pupil constructed as expert, encouraged to apply science to all aspects of everyday life, with activity in science classroom as weighted heavily toward inquiry processes. In the midst of this strong institutional pull, Young wonders: Shouldn't disciplinary knowledge and thinking matter? Don't we want to insure that students are exposed through the school curriculum to "better," "more reliable" knowledge, that which is "nearer to truth about the world we live in and what it is to be human" (2013: 107)? He worries that students, especially those from marginalized groups, are not given opportunities to learn the specialized knowledge that they do not have access to at home. He indicts curriculum theorists for failing to address these sorts of questions. Deng (2015) responds that the Didaktik tradition offers a potentially useful model of human development through "encounters" with academic content, that is, development *through* disciplines. This foundation allows curriculum making that can "discover and appraise disciplinary content... in terms of its educational potential for fostering intellectual and moral powers."

Yet, neither Young nor Deng contend with the fact of a rapidly changing technical environment that is likely to shape schooling in profoundly new ways, despite the strength of the institutional environment that appears to discourage the teaching and learning of Young's "powerful knowledge" at least under some circumstances. The fundamental change in the technical environment shaping schooling is the shift from information scarcity to information abundance (Data, data everywhere 2010; Eisenberg 2008, Mayer-Schönberger and Cukier 2013). Whereas the key problem in applying knowledge in the past was its scarcity, with schools being a place to access this scarce resource, developments in ICT have created the opposite problem: finding appropriate knowledge to solve a particular problem from the ocean of information available. The internet has impacted what Eisenberg (2008) called the "landscape of information." Since then, however, given that everyone, even school-aged children, can be not only gatherers but also producers and transmitters of information using ICT, a more apt characterization is that ICT supports an entirely new "ecosystem of knowledge" that curriculum makers need to acknowledge and take into account.

Admittedly, the internet is clearly a repository for both the everyday knowledge gained from everyday experience – memes, developments in popular culture, the sharing of news and opinions by extended social networks – and the theoretical/abstract disciplinary knowledge that Young wants students to access through schooling (McEneaney 2015). The challenge, of course, is for internet users to discern when data and information found on the internet is, in fact, *knowledge* that can be applied to a particular decision-making situation, since everyday knowledge is jumbled together with the more reliable discipline-based knowledge. This is a challenge taken up by advocates of including "information literacy" in the school curriculum (Warschauer 2007, Riedling 2007).

In fact, however, the major jolt to the institutionalized curriculum lies in the fact that what counts as "everyday" for a child has changed dramatically through this shift in the technical environment, and therefore what counts as "relevant" in primary science curricula is shifting too. Any child on the planet with a smartphone and the free Google Earth app can tour the Pyramids of Egypt or volcanoes on Mars. Toddlers watch educational videos, while their parents do the weekly grocery shopping. Weather Underground invites lay people to connect personal weather stations to contribute local weather data via the company website. Developments in virtual reality technology and data visualization promise to make these everyday experiences in this new ecosystem of knowledge more immersive. In short, the internet has rapidly become a place, not so much to "visit" to gather information (Web 1.0) but to *be*, to seek and have experiences, particularly for school-aged children (Web 2.0 and counting) (Greenhow et al. 2009). Increasingly, too, access to this new ecosystem of knowledge is becoming taken for granted around the world, with smartphone ownership by adults in emerging and developing countries nearly doubling, from a median of 21% to 37%, between 2013 and 2015. Although people in wealthier countries still use the internet more and own more internet-enabled devices, a recent study concludes "the rest of the emerging world is catching up" (Pew Research Center 2016).

Consequently, the direct challenge exists virtually everywhere to schools and school curricula. One response is to try to seal off schools from the demands of this changed technical environment, essentially walling off pupils from this new ecological system of knowledge, by very tightly controlling student access to the internet during the school day. Even with more liberal access to ICT, challenges persist. Beautifully teacher-crafted inquiry activities intended to be collaborative, realistic, and relevant will not be: Why wouldn't students search the internet or post on social media to find reasonable approaches to the problem posed, if not the "correct" answer? And we need to ask ourselves: Is that use of ICT resources a bad thing?

Will pupils invest persistent effort in their small collaborative work groups when a YouTube video might explain an idea much better than their group can co-construct it? On a more basic level, what disciplinary terms does one need to know when an authoritative dictionary is a click away? Does a primary science pupil ever need to memorize local flora and fauna when one can snap a picture and an app will reveal the basics, while also providing links to more abstract disciplinary knowledge if desired? To what extent are school curricula, and particularly primary science curricula, preparing teachers to teach and children to learn from these internet-based experiences, and life in this new ecosystem of knowledge? In the next section, I consider which of the East Asian systems reviewed here seem to be better positioned to address the challenges posed by this radically changed technical environment.

13.4 Supports and Challenges to Science Learning in the New Ecology of Knowledge

There is much reason for optimism that the systems and strategies in place in East Asia as described in this volume can be supportive of learning amidst this jarring change.

Views of School Science Knowledge In many ways, the long-term institutionalized pull toward emphasizing process skills versus discrete content positions school systems to embrace ICT resources that provide ready access to content. It could be argued, in fact, that ICT itself would not have developed as quickly and been demanded by so many, without this prior shift toward process and without the democratization of expertise that had been brewing for decades. Thus, Taiwan's aim to "cultivate science concepts" is likely to work well in an information-abundant landscape, as is Korea's focus on skill-based "core competencies" balanced with "understanding basic concepts." Highlighting systems thinking and "big ideas" while experimenting with learning objectives that are intentionally not specific suggests that Japan's view of primary school science knowledge is ready for exploring through digital means, as is setting aside time in the Period for Integrated Studies for pupils to "think in their own way about life." The Korean aim to develop "interdisciplinary human talent" comports with the wide-open nature of knowledge on the internet. In contrast, recent research results reported in this volume show that Singapore's learning objectives for primary science heavily emphasize the lower levels of Bloom's revised taxonomy (with "remembering" and "understanding" levels comprising nearly two-thirds of all learning objectives), deviating from institutionalized norms that lean toward the "content-free."[2] However, this implies that Singapore's intended and perhaps also its enacted primary science curricula are not primed to accommodate the challenges posed by ICT.

Boundedness of Science Learning Another broad principle that may determine how school systems deal with these changes in the technical environment is the extent to which learning is expected to happen "in school" or whether boundaries between in-school and out-of-school learning have been softened. Systems that invite and value learning outside of the physical confines of school are likely to transition more easily to embrace learning in the virtual sphere. Thus, science competitions and camps, such as those common in Singapore, China, and elsewhere, bode well for effective adaptation to ICT realities, as does the Chinese principle that science curriculum should be flexibly constructed, with not only schools, but "family, society, and a variety of other resources" expected to enrich the learning experience. Seen from this perspective, Korean primary science teachers' tendency to rely heavily on a textbook undercuts the notion, essential for mining the promise of ICT, that valuable knowledge could be found and experienced in non-school locations. Hong Kong's explicit embrace of information technology, its reported commitment to a "resource-based learning environment" including multimedia, and efforts to avoid "teacher dominance" in the use of internet-based resources hint at a promising "open source," nonhierarchical, stance toward school knowledge.

[2] Traditional sociological perspectives consider not only the sources of conformity and isomorphism but also why certain actors are empowered to deviate from normative structures. Perhaps Singapore's decades-long position at the top of TIMSS rankings in primary science and math engenders legitimacy that allows some latitude for deviance of this kind. Focus on lower-level reasoning in the curriculum may also boost mean scores on the international tests, which continue to be weighted toward recall and basic understanding. This dynamic contributes, in part, to the assessment "bottleneck" described in the report from Singapore in this book.

Framing Policy Documents Applications of institutional theory in education, health care, human rights, and other areas suggest that policy documents providing language that frames the issues and acceptable actors and solutions can shape organizational responses. Often, these policy documents, including treaties, high-level reports, and statements of standards or regulations come from international governmental and nongovernmental organizations (e.g., United Nations Convention on the Rights of the Child, Education for All, ISO management system standards). Such framing documents might merely articulate and reify elements in the institutional environment, but they might also support more nimble responses to the technical environment. In this case, reports or strategic plans produced within nations might provide discursive resources that shape orientations and provide ready justifications for investments that might usher in faster adjustments to the ubiquity of ICT. Singapore's third Masterplan for ICT in Education, for example, not only calls for improved technology in schools but also discursively outlines a new direction for curriculum and pedagogy by aiming to "strengthen competencies for self-directed learning" and "enable students to learn anywhere" (Infocomm Media Development Authority 2016). An early framing document in Japan, the Central Council for Education 2005 Report on the "knowledge-based society," also seems to discursively prime the system for more effectively embracing the new ecosystem of knowledge, as it emphasizes that knowledge is borderless as it progresses rapidly and with frequent paradigm shifts. The necessity of flexible cognitive skills, engaging constantly with new flows of information as on the internet, follows from this view of the knowledge-based society.

Information Literacy As pupils increasingly are immersed in ICT as part of their school activities, information literacy in some form should be part of the regular curriculum to help them discern reliable, discipline-based knowledge from the more ephemeral everyday knowledge or even the unashamedly "fake news." In the United States, most of this effort has been at the secondary and college and university levels (Riedling 2007), but it seems clear that there is a need to introduce these skills at a younger age. Currently, a number of East Asian countries (e.g., Taiwan, Singapore) include critical thinking as part of the primary science curriculum, but I would argue that information literacy requires specialized knowledge and skills that build on basic critical thinking. One way to be a more astute consumer of science information is to hone one's own science communication skills, in order to recognize rigor, precision, and unbiasedness in others' written, oral, or visual communication about science. In this, Korea seems to be on the forefront. Taiwan's new primary science curriculum, to be implemented in 2018, is very promising in this regard as it requires pupils to reflect on science reported in the media and emphasizes science reading and learning to access "public science." The push for a "public science literacy" in China's curriculum would also appear to require some attention to information literacy skills.

Assessment Strategies Many of the accounts of primary science in this book highlight the influence of assessment systems on curriculum and pedagogy. When it comes to preparing children for a life in this new ecosystem of knowledge, some of the existing assessment systems seem more amenable to achieving this goal. Japan's national assessment poses authentic problems grounded in everyday experiences that are not Google-able; however, the closed-ended nature of the assessment is less supportive of this aim, and Taiwan's "Hands On, Minds On" primary science is hampered by similar assessment problems. The performance-based assessment that Korea is developing could potentially incorporate ICT resources in ways that are authentic while emphasizing skills over easily searchable content. Pupils in China are asked to apply knowledge and solve problems, and the expectation, reported in this book, that they would "[process] a variety of information to access facts" during assessment activities implies a curricular orientation that pushes students to confront and eventually cope with the relatively new technical problem of information abundance.

Teachers as Members of the Science Community In advocating for more emphasis on "powerful knowledge" in the school curriculum, Young (2013) highlights the particular importance of drawing on the structure of disciplinary knowledge and connections to disciplinary communities themselves. At the primary level, ICT resources can help teachers, who also spend much of their lives outside of work online and using social media and other aspects of ICT, continue to learn about science disciplines, as well as connect to the broader science community. Some national systems are more forward leaning than other systems by specifically incorporating ICT to support teachers as learners of science and in their development of science pedagogy. In Taiwan, primary science teachers are reported to often forge ties with other teachers through various Facebook groups, sharing ideas and resources. ICT was cited as supporting teacher learning as well in Hong Kong, whether that involves online videos or participation on a variety of discussion boards. In the future, I expect more harvesting of the potential for teachers' interactive learning using ICT resources, including more immersive learning experiences through virtual reality-based applications.

Teachers, like their pupils, could probably benefit from sharpened information literacy skills, as they develop horizontal virtual linkages through social media with other teachers, but ICT holds great potential for helping teachers develop what might be characterized as "vertical" linkages with professional science and scientists. Many professional science associations, such as the American Physics Society, already host websites with Web 2.0 interactive possibilities, such as posting questions or comments about a research brief. As teachers grow in their own knowledge of science and science teaching, this expertise will help them effectively curate sets of internet resources (rather than select a single resource) as a means to scaffold pupils' work online, while approximating an authentic, information-abundant, ecology of knowledge.

13.5 Conclusion

In this chapter, I have highlighted the many similarities that characterize primary school science in the six East Asian systems described in this book, and I argue that institutional perspectives that highlight the models and schema transmitted through a "world culture" can explain this isomorphism. This is not to say that there are no differences between accounts nor differences between these East Asian approaches and those of other regions in the world. For example, the fact that so many East Asian countries employ specialist science teachers at the elementary level would be surprising to many North American readers, where primary teachers are charged with teaching all subjects (though perhaps supported at the school or district level by a science curriculum coach). Several of the accounts in this book mentioned ethics as a topic, and that is not yet an institutionalized universal in the primary science curriculum. Nonetheless, the richly detailed accounts from the five systems provide ample evidence of the strength of the institutional environment.

Yet organizations, including national systems of schooling, are shaped by the technical environment as well. Until fairly recently, the technical environment's influence on mass schooling has been relatively weak: the research base was not solid enough to delineate clear means-to-ends processes for learning, and there had been little in the way of truly game-changing technological innovations in school settings. I contend that the rapid development and especially widespread, 24-h access to Information and communication technology (ICT) is already transforming the technical environment to which mass education systems must respond. In particular, ICT means that children (and their teachers) are inhabitants of a new ecosystem of knowledge and that curricula and pedagogical strategies urgently need to accommodate to this new reality of information abundance rather than scarcity. Primary science formulations that have embraced process skills over discrete content, that see learning as happening in and out of school, and that support pupils to sift through content wisely using information literacy skills are best positioned to draw fruitfully on ICT resources to education children in powerful, engaging ways. What this profound change in the technical environment of mass schooling means for the influence of the institutional environment is not yet known.

References

Baker, D. (2014). *The schooled society: The educational transformation of global culture*. Stanford: Stanford University Press.

Baker, D. P. (2015). A note on knowledge in the schooled society: Towards an end to the crisis in curriculum theory. *Journal of Curriculum Studies, 47*(6), 763–772.

Bereiter, C., & Scardamalia, M. (1996). Rethinking learning. In D. R. Olson & N. Torrance (Eds.), *The handbook of education and human development: New models of learning, teaching and schooling* (pp. 485–513). Cambridge, MA: Basil Blackwell.

Boli, J., & Thomas, G. M. (1999). *Constructing world culture: International nongovernmental organizations since 1875*. Stanford: Stanford University Press.

Data, data everywhere. (2010, February 27). *The Economist*, p. 3.

Deng, Z. (2015). Content, Joseph Schwab and German Didaktik. *Journal of Curriculum Studies, 47*(6), 773–786.

Eisenberg, M. B. (2008). Information literacy: Essential skills for the information age. *DESIDOC Journal of Library & Information Technology, 28*(2), 39.

Frank, D. J., & Gabler, J. (2006). *Reconstructing the university: Worldwide shifts in academia in the 20th century*. Stanford: Stanford University Press.

Greenhow, C., Robelia, B., & Hughes, J. E. (2009). Learning, teaching, and scholarship in a digital age Web 2.0 and classroom research. *Educational Researcher, 38*(4), 246–259.

Infocomm Media Development Authority. (2016, November). *Singapore unveils third Masterplan for ICT in Education*. Retrieved from: https://www.imda.gov.sg/infocomm-and-media-news/buzz-central/2008/9/singapore-unveils-third-masterplan-for-ict-in-education

Lee, Y.-J., Kim, M., & Yoon, H.-G. (2015). The intellectual demands of the intended primary science curriculum in Korea and Singapore: An analysis based on revised Bloom's taxonomy. *International Journal of Science Education, 37*, 2193–2213.

Mayer-Schönberger, V., & Cukier, K. (2013). *Big data: A revolution that will transform how we live, work, and think*. New York/Houghton Mifflin: Harcourt.

McEneaney, E. H. (2003a). Elements of a contemporary primary school science. In G. S. Drori, J. W. Meyer, F. O. Ramirez, & E. Schofer (Eds.), *Science in the modern world polity: Institutionalization and globalization* (pp. 136–154). Stanford: Stanford University Press.

McEneaney, E. H. (2003b). The worldwide cachet of scientific literacy. *Comparative Education Review, 47*(2), 217–237.

McEneaney, E. H. (2015). Finding knowledge on the internet: Implications for the knowledge-driven curriculum. *Journal of Curriculum Studies, 47*(6), 802–819.

McEneaney, E. H., & Meyer, J. W. (2000). The content of the curriculum: An institutionalist perspective. In M. T. Hallinan (Ed.), *Handbook of sociology of education* (pp. 189–211). New York: Kluwer Academic/Plenum Publishers.

Meyer, J. W., & Scott, W. R. (1983). *Organizational environments: Ritual and rationality*. Beverly Hills: Sage.

Meyer, J. W., Boli, J., Thomas, G. M., & Ramirez, F. O. (1997). World society and the nation-state. *American Journal of Sociology, 103*(1), 144–181.

Meyer, J. W., Kamens, D., & Benavot, A. (2017). *School knowledge for the masses: World models and national primary curricular categories in the twentieth century* (Vol. 36). London: Routledge.

Morrow, W. (2008). *Bounds of democracy: Epistemological access in higher education*. Pretoria: HSRC Press.

Pew Research Center. (2016, February). *Smartphone ownership and internet usage continues to climb in emerging economies*. Retrieved from: www.pewresearch.org

Popkewitz, T. S. (1997). The production of reason and power: Curriculum history and intellectual traditions. *Journal of Curriculum Studies, 29*(2), 131–164.

Riedling, A. M. (2007). *An educator's guide to information literacy: What every high school senior needs to know*. Westport: Libraries Unlimited.

Scardamalia, M., & Bereiter, C. (1996). Engaging students in a knowledge society. *Educational Leadership, 54*(3), 6–10.

Schofer, E., Ramirez, F. O., & Meyer, J. W. (2000). The effects of science on national economic development, 1970 to 1990. *American Sociological Review, 65*(6), 866–887.

Scott, W. R. (2003). *Organizations: Rational, natural and open systems* (5th ed.). Upper Saddle River: Prentice-Hall.

Scott, W. R., Ruef, M., Mendel, P. J., & Caronna, C. A. (2000). *Institutional change and healthcare organizations: From professional dominance to managed care*. Chicago: University of Chicago Press.

Warschauer, M. (2007). Information literacy in the laptop classroom. *Teachers College Record, 109*(11), 2511–2540.

Young, M. (2008). *Bringing knowledge back in*. London: Routledge.

Young, M. (2013). Overcoming the crisis in curriculum theory: A knowledge-based approach. *Journal of Curriculum Studies, 45*(2), 101–118.

Chapter 14
Coda

Yew-Jin Lee and Jason Tan

> The most useful comparisons, it seems, are those that identify the common practices of the high-performing countries and systems. Notwithstanding the socio-cultural, economic and other significant differences that may become very evident during the comparison process, a rigorous methodology should extract the useful (and potentially transferable) conceptual as well as practical approaches that underpin education in general, and curriculum design in particular. This requires a nuanced methodology that enables consideration of local characteristics and objectives while maintaining an openness towards a wide range of international practices. *If comparative education is to assist reform, then it must be a tool for both inquiry and analysis.* (ACARA 2017, p. 8, italics our emphasis)

It is extremely difficult to either act as a spokesperson or fortune-teller for one region much less attempt to speak on behalf of all six regions that have evolved so differently and celebrate that very fact! We felt that distilling (and arguably misrepresenting to some extent) some of the possible next steps for primary science education in East Asia from our expert commentators in Part II could open up a few preliminary insights before we furnish some thoughts of our own. Hence, what we see below is a list of their chief recommendations, questions, or jumping-off points for discussion though actual engagement with their writing is really nonnegotiable:

(a) Lefstein – To reconsider forms of scientific literacy that are mindful of human meaning, purpose, and solidarity for democratic citizenry.
(b) Russell – Having well-trained teachers using ambitious/productive pedagogies to nurture young people to think and reason like scientists.
(c) Ramnarain – How do we adequately support teachers who are moving from a curriculum that previously valued content (i.e., knowledge as product) toward one that emphasizes inquiry (i.e., knowledge as process)?

Y.-J. Lee (✉) · J. Tan
National Institute of Education, Nanyang Technological University, Singapore, Singapore
e-mail: yewjin.lee@nie.edu.sg

Y.-J. Lee, J. Tan (eds.), *Primary Science Education in East Asia*, Contemporary Trends and Issues in Science Education 47,
https://doi.org/10.1007/978-3-319-97167-4_14

(d) Symaco and Daniel – Planning instruction in science that is responsive to rapid changes in an increasingly unpredictable world (e.g., due to technology, market forces, social changes).
(e) Zembal-Saul – Seek greater coherence within and between educational policies, assessment, curricula, teacher preparation, and other domains to achieve meaningful learning of science via inquiry.
(f) McEneaney – Strong institutional environments here have to articulate with the numerous transformations in the technical environment that have implications for the location, quantity, type, and generation of (new) knowledge for youth.

Each of the above ideas or expressions of hope represent the consolidated wisdom from our experts and to which we too largely agree with. As coeditors, we now offer a few tentative possibilities and visions for the road ahead without forgetting that predicting trends is a terribly inexact activity. First, we are fairly certain that because of the intensified diffusion of ideas through outlets like journals (c.f. Lee 2015), conferences, multidirectional academic exchanges across continents, and international large-scale assessments, East Asian systems are likely to experience even greater converge in their strategies and goals. All of these resonate with the changing DNA of education that regards issues of accountability, standardization, and performativity as normative. One piece of evidence that astonished us here was that Japan, which has an elementary science curriculum that is most divergent from other East Asian regions (see Lee et al. 2017) has acknowledged that reactions to PISA testing has already penetrated policy and assessment practices in that country (Matsubara, Chap. 3 this volume)!

Moreover, the magnetic attraction that major curriculum developments in the United States (e.g., the Next Generation Science Standards) have on East Asian policymaking cannot be underestimated as Asian educators search for new directions (fads?) to enhance their own education systems. So if a second edition of this book were to be published within the next 8 years, we would not be in the least bit surprised to find notions such as scientific practices, cross-cutting ideas, and core concepts being pushed in textbooks. While this might result in a certain uniform blandness or monotony with respect to curriculum making and pedagogic strategies, we are nonetheless very pleased if the kinds of key epistemic skills so explicitly advocated in the NGSS were to be more widely available for students here. This is both their entitlement and need as young people move further into the twenty-first century that is growing in sociotechnical and ecological uncertainties. And given how education systems in Asia pay attention to align science teacher preparation and PD to serve their goals, we do anticipate major challenges ahead if past experience with the promotion of inquiry learning in schools is anything to go by. At primary grades, asking for such elevated forms of science instruction from teachers who themselves might not have had the requisite background in science nor are comfortable with recontextualizing scientific concepts for young children exacerbates such policy practice disjunctures immensely. Nonetheless, we do also expect to see more confident examples of curriculum leadership and innovation within these six East-Asian regions as their educational systems mature.

Second, we hope that we can realize a more equitable and just elementary science education in East Asia. With a reputation for being merely hardworking and exam-focused, there are many learners here who do not subscribe to this kind of work ethic or who struggle to find any meaning associated with studying science. And we should not forget students with special needs or those who are tracked or pulled out of mainstream education such as for reasons of poverty, ignorance, or discrimination. We do not deny that there are individuals who move at a slower pace of readiness, but we do want to be able to assure all learners access to inclusive forms of science education although this might be manifested differently across the various regions. As we have mentioned in the introductory chapter, there are indeed many "Asias" to be found depending on where and how one chooses to look. These are some of the contentious questions concerning the purposes and meaning of (science) education that have been explored in many ways throughout our book (see Biesta 2009).

We argue that the trajectories of student learning in primary science classrooms need to be understood more clearly in our third interrelated point. In much of Asia, teacher authority is a given and not to be debated, but we believe there is room for learners here to assume greater agency and develop science identities and self-actualization without compromising respect for their teachers. These issues also overlap with the tensions between enjoyment in learning and learning as struggle although this is not a strict dichotomy as commonly believed. So if we are serious about examining what is going on in elementary science classrooms and the interactions between teachers and students, then pioneering international projects such as the OECD Teaching and Learning International Survey (TALIS) Video Study can mobilize interest and provide recommendations for change to the enacted curriculum, which was only tangentially covered here and perhaps rightly so. Most problems related to student learning and achievement are located *during* the implementation of curricula rather than with the nature or organization of institutional curricula/systems, the primary focus of this present volume. Knowing the former is therefore vital to understand the routine accomplishment of primary science lessons and will dialectically strengthen the backbone of an evidence-based methodology in comparative curriculum/policy research as shown in our opening quote. The practical difficulties of this task are, however, not to be taken lightly for "[a]ny description of pedagogy would be incomplete if the mediation of teachers' actions by the surrounding culture is not taken into consideration" (Zyngier 2016, p. 178). Researchers of elementary science education in East Asia truly have their work cut out for them for a long time!

References

Australian Curriculum and Assessment Authority (ACARA). (2017). *Literature review: Contemporary approaches to comparative education research.* Retrieved from https://www.australiancurriculum.edu.au/resources-and-publications/publications/program-of-research-2017-2020/

Biesta, G. (2009). Good education in an age of measurement: On the need to reconnect with the question of purpose in education. *Educational Assessment, Evaluation and Accountability, 21*, 33–46.

Lee, Y.-J. (2015). In a different voice: promises and trials of non-English medium journals. *Cultural Studies of Science Education, 10*, 1051–1055.

Lee, Y.-J., Kim, M., Jin, Q., Yoon, H.-G., & Matsubara, K. (2017). *East-Asian primary science curricula: An overview using revised Bloom's Taxonomy*. Dordrecht: Springer.

Zyngier, D. (2016). Culture and pedagogy(ies): (What) have learned from and since Alexander 2001. In D. Wyse, L. Hayward, & J. Parker (Eds.), *The SAGE handbook of curriculum, pedagogy and assessment* (pp. 171–186). London: Sage.

FSC
www.fsc.org
MIX
Papier | Fördert
gute Waldnutzung
FSC® C083411

Kolibri 360 GmbH
Ferdinand-Jühlke-Straße 7
99095 Erfurt, Deutschland
produktsicherheit@kolibri360.de